"Until now, eco-tourists visiting the Great Plains faced a void of geological information. . . . [*Great Plains Geology*] is clearly and succinctly written by a leading geologist in a way that nongeologists will understand and appreciate."

—JAMES STUBBENDIECK, director emeritus of the Center for Great Plains Studies at the University of Nebraska–Lincoln and author of *North American Wildland Plants*

"Professor Diffendal has done a marvelous job of assembling information and images about the rich geological history and terrain of the Great Plains. For those who have ever lived in or spent time in the region, as I have, or been as smitten with geology as I was. . . . I highly recommend it."

—ROBERT WUTHNOW, professor of sociology at Princeton University and author of *Remaking the Heartland: Middle America since the 1950s*

"An enjoyable guide to the best geological sites in the Great Plains of Canada and the United States. Professor Diffendal's lively writing unites geology with personal and historical references to provide a great resource for those traveling and sightseeing."

—DAVID WATKINS, professor of earth and atmospheric sciences at the University of Nebraska–Lincoln

"From the inquisitive tourist or landowner, to the 'wannabe' archaeologist or dinosaur hunter, to the professional scientist or historian who seeks information in a related field, this book is a *must-read*. . . . [It] will quickly dispel the idea that the Great Plains are a monotonous and continuously flat region."

—GERALD SCHULTZ, professor of geology at West Texas A&M University

GREAT

DISCOVER THE GREAT PLAINS

Series editor: Richard Edwards, Center for Great Plains Studies

R. F. DIFFENDAL JR.

PLAINS

Geology

UNIVERSITY OF NEBRASKA PRESS   *Lincoln and London*

A Project of the Center for Great Plains Studies, University of Nebraska

Library of Congress
Cataloging-in-Publication Data
Names: Diffendal, R. F. (Robert Francis)
Title: Great Plains geology / R. F. Diffendal Jr.
Description: Lincoln: University of Nebraska
Press, [2017] | Series: Discover the Great Plains
| Includes bibliographical references and index.
Identifiers: LCCN 2016034809 (print)
LCCN 2016038427 (ebook)
ISBN 9780803249516 (pbk.: alk. paper)
ISBN 9781496200778 (epub)
ISBN 9781496200785 (mobi)
ISBN 9781496200792 (pdf)
Subjects: LCSH: Geology—Great Plains. |
Physical geography—Great Plains. | Great Plains.
Classification: LCC QE78.7 .D54 2017 (print)
| LCC QE78.7 (ebook) | DDC 557.8—dc23
LC record available at
https://lccn.loc.gov/2016034809

Set in Garamond Premier by Rachel Gould.
Designed by N. Putens.

To my parents, who supported my undergraduate training, to Jacob Freedman of Franklin and Marshall College, who inspired me to further my education, to J. A. Fagerstrom, who nurtured me through graduate work, and especially to my wife, Anne, who has always been ready to accompany me on our adventures and to improve my manuscripts about my research and travels.

CONTENTS

ILLUSTRATIONS

PREFACE

This book offers my personal view of the geology of the Great Plains. It is intended for ecotourists, anyone with a broad interest in geology and some general education in science, professional geologists and geographers wanting to become more familiar with the region, and students, farmers, ranchers, and K-12 educators who want to know about the Great Plains and its geological development. Any errors of interpretation or fact in this book are solely my own.

The book is arranged in three chapters. The first two discuss the region's boundaries and provide a brief geologic history to explain how the Great Plains came to be in its present form. The third chapter describes 57 outstanding geologic sites from Alberta and Saskatchewan, Canada, to southern Texas. Features at these sites illustrate various aspects of the geology of the Great Plains. Most of the photographs of these sites are my own.

Figure 1, the geologic time scale and rock column, follows the acknowledgments, not because you need to refer to it now, but for you to note it for later. You will find it helpful throughout this book. Use this figure to see the chronological relationships among the geological time spans and the rock units deposited in those time spans. The names of these time spans and rock units appear frequently in the book. Use this figure also to identify the occurrence of important events that affected the Earth's development, such as mass extinctions, glaciations and meteor impacts.

I have included three appendices for background: the first on the boundaries and subdivisions of the Great Plains; the second an overview of the development of key geologic concepts; and the third a list of cautions regarding travel in the Great Plains. These are followed by a glossary of the geologic terms used in the text, a list of references, and an index.

For easier reading, I have eliminated the accepted scientific style of citing references in the text in favor of a simplified style of citing the name of the principal author of a work on some of the topics I am discussing. Please refer to that name in the references section at the end of the book to find the appropriate source.

ACKNOWLEDGMENTS

I am indebted to Richard C. (Rick) Edwards, director of the Center for Great Plains Studies at the University of Nebraska, for providing me with office space, hardware, and supplies during my research and writing. Financial support for travel and associated activities came from the University of Nebraska–Lincoln Emeriti Association and the Conservation and Survey Division of the School of Natural Resources, University of Nebraska–Lincoln. Ms. Dee Ebbeka drafted several of the figures.

To my colleagues and friends who tramped with me and talked with me about the geology of the Great Plains over the years—Mike Voorhies, Bob Hunt, Jim Swinehart, George Corner, Larry Agenbroad, Cinda Temperly, Gerry Schultz, Tex Reeves, Tom Gustavson, Dale Winkler, Jason (Jay) Lillegraven, and others—I express my gratitude for all of their help.

Thanks to Professor Wakefield Dort (University of Kansas), Shane Tucker (University of Nebraska State Museum), Kim Weide, Sheryl Cooley, Heather Lundine, Michael Forsberg, and Rick Edwards for their helpful comments that led to a very much improved manuscript. I especially thank my wife, Anne; Bridget Barry, Emily Wendell, Joeth Zucco, and other staff at the University of Nebraska Press; and the copyeditor, Joy Margheim, for offering very helpful suggestions on additions and reorganizations of several drafts of the manuscript that made it much more readable than in its original form.

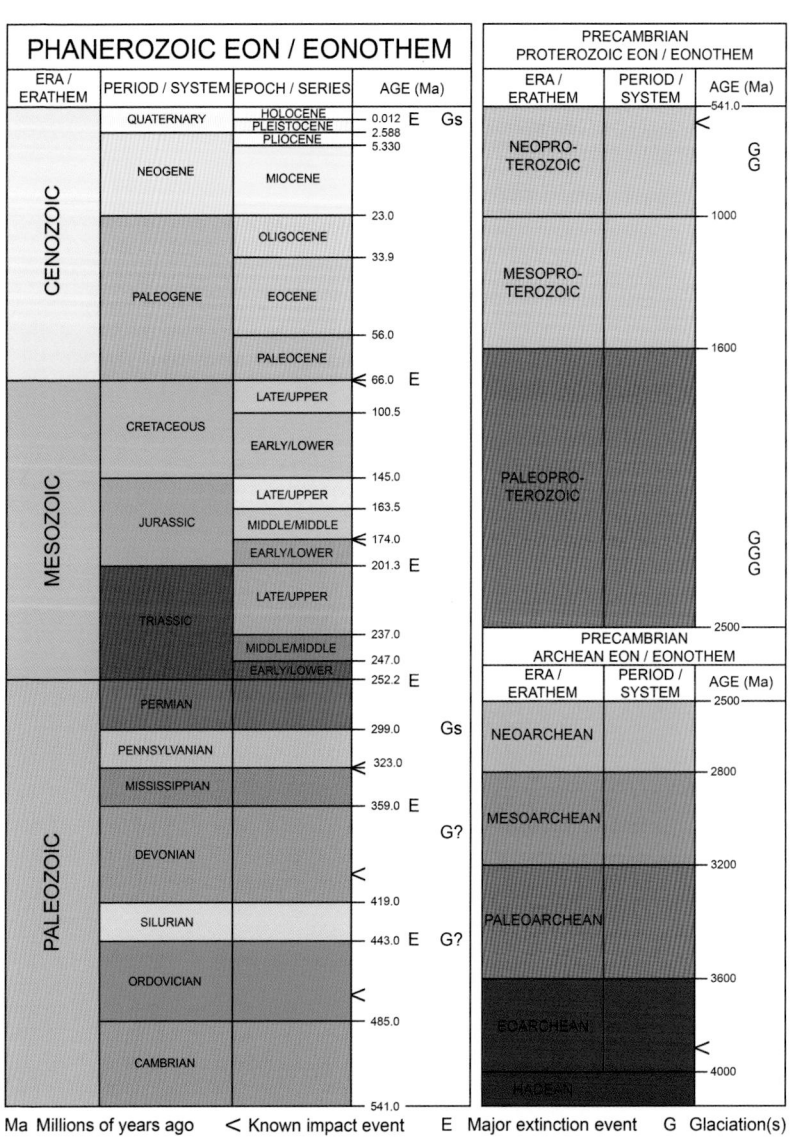

Fig. 1. Geologic time scale and rock column. U.S. Geological Survey.

INTRODUCTION

Like art and beauty, what constitutes the Great Plains lies in the eye of the beholder. For people who have never traveled in the American heartland, the name may conjure up images of cowboys, Indians, buffalo, prairie dogs, grasshopper swarms, cattle drives, heat, wind, dust, blizzards, tornadoes, floods, and very flat land. For those who live in the central parts of the lower 48 states, it can be some or all of these things. After hearing a college student of mine from the inner city of Chicago define a prairie as a vacant city lot, I came to better understand that our views are shaped by our experiences and differ from one another's perceptions, often in major ways. "I know what the Great Plains region is, and this place is clearly not part of it!" one person would say and another would differ completely.

Captain William Clark thought that he saw where the plains began on the west bank as he traveled up the Missouri River near the present-day border of Kansas and Nebraska on July 10, 1804. Others have seen the beginning of the Great Plains in places to the east or to the west of that place (fig. 2).

John Wesley Powell, ethnographer, writer, explorer of the American West and second director of the U.S. Geological Survey, was, so far as I know, the first person to describe and to draw a map showing the boundaries of the major physiographic regions of the lower 48 states, including the Great Plains (fig. 3). Powell also recognized that the Great Plains extended northward into Canada and southward into the Republic of Mexico. Later published works have varied on the size and boundaries of the Great Plains, but Powell was probably the first scholar to give the region that written name and a description. And that name has stuck both in literature and in the popular imagination.

For me, the Great Plains is a land of wide-open spaces with few trees except along watercourses or on local uplifts like the Black Hills. It is a land of big, wild, and scenic rivers, sadly now controlled to a greater or lesser extent by dams and water diversions. It is a land usually short of water, yet it includes one of the major aquifers (variously called the High Plains, Ogallala, or High Plains/Ogallala; hereafter the High Plains aquifer) in the United States. The Great Plains can look markedly different at any one place, depending on the time of day and the quality of the light; there is a mirage-like quality to the country. Plain or otherwise, it is a beautiful place, often of few people and much livestock. I love the region.

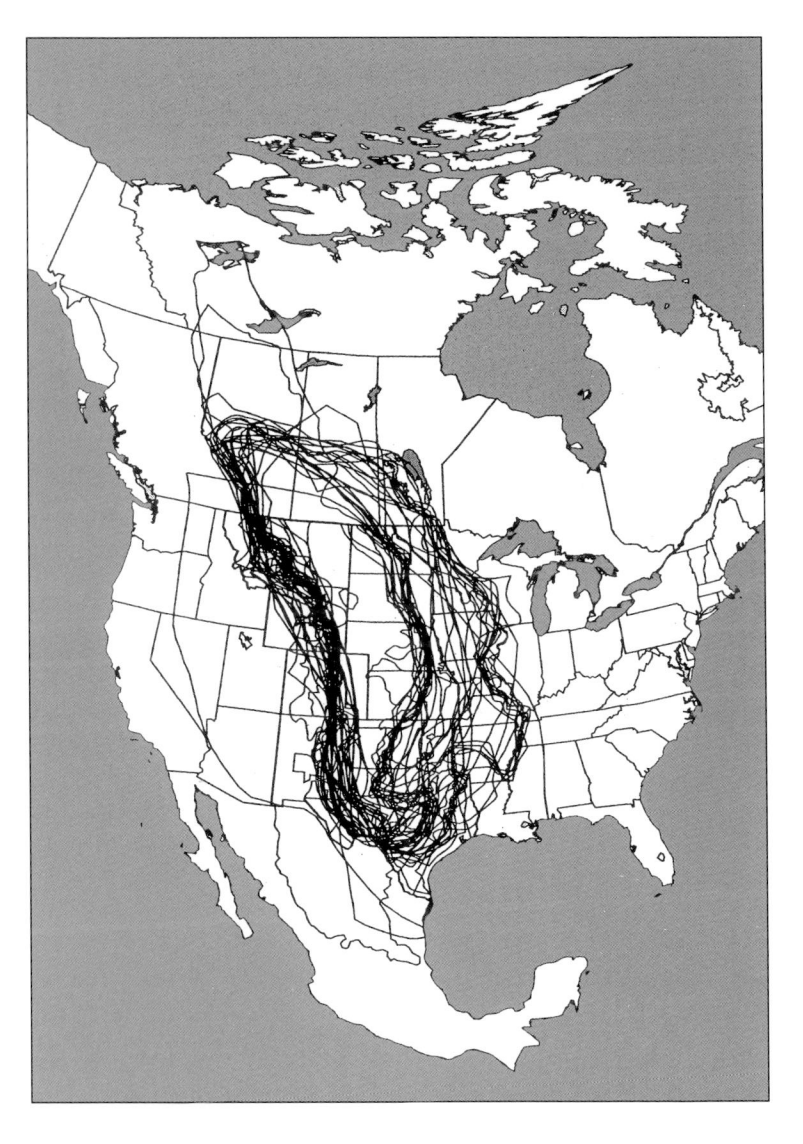

Fig. 2. Fifty published versions of the Great Plains regional boundary. Copyright 2000. Reproduced from figure 1, p. 546, of "Where Are the Great Plains? A Cartographic Analysis," *Professional Geographer* 52, no. 3, by Sonja Rossum and Stephen Lavin. Reproduced by permission of Taylor & Francis Group, LLC (http://www.tandfonline.com).

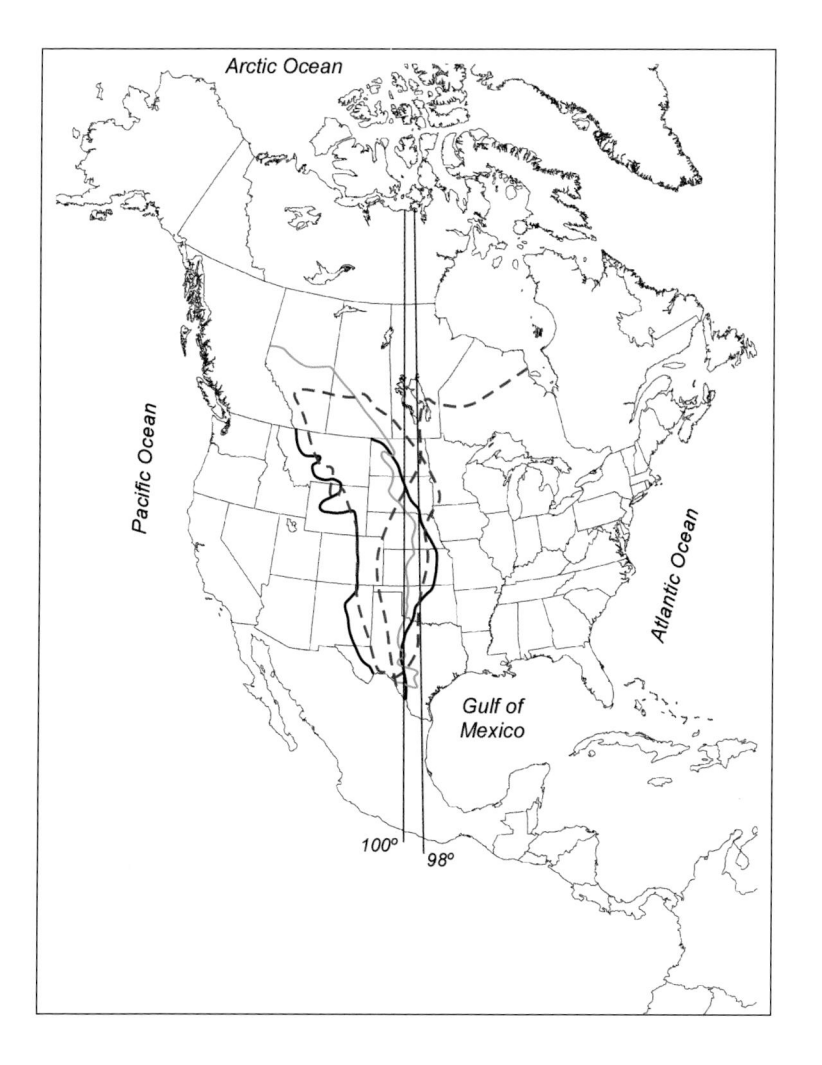

Fig. 3. J. W. Powell's boundaries for the Great Plains in the United States and some of the other possible choices proposed for the eastern boundary. The green line is the approximate 2,000-ft. contour; the blue dashed line is the approximate 20-in. precipitation line; the red line is the approximate area of alkaline and transitional soils. Solid black lines are Powell's eastern and western boundaries of the Great Plains. From Powell, "Physiographic Regions," 98–99.

GREAT PLAINS GEOLOGY

CHAPTER 1

What Is the Great Plains?

Drawing the Boundaries

With the exception of the western edge, bordered mostly by the Rocky Mountains, the Great Plains is a region (also called a physiographic province) with few obvious boundaries. In the years since John Wesley Powell first mapped, described, and named it, few authors have agreed where the Great Plains begins or ends. In contrast, most of the other physiographic provinces of the United States, Canada, and Mexico are rather sharply delineated. Fifty published maps of the region show many different versions of its supposed boundaries (fig. 2). Some include land as far north as the Canadian Northwest Territories, as far south as Mexico, as far east as Illinois, or as far west as Utah. Some more recent ones have even more unusual boundaries, including one that puts Indiana, Michigan, and Ohio in the Great Plains but leaves out any states south and southwest of Kansas. Almost all have boundaries drawn as solid lines, indicating a certainty that is clearly not so. In fact, all are approximations based on either the understanding or the misunderstanding of the individual map maker.

To me, the extremes are not parts of the Great Plains. Most of the western boundary lines shown in figure 2 cluster along the break between the plains and the eastern slopes of the Rocky Mountains. The northern boundaries lie mostly in the southern parts of the Canadian Prairie Provinces, and the southern boundaries lie mostly in western Texas. The eastern boundaries are more problematic. However, two clusters of boundary lines are recognizable. One of these clusters mostly follows the upper Mississippi Valley; the other lies to the west of that some hundreds of miles.

Drawing a boundary for the Great Plains is no easy task for someone who wants to be precise. To make matters worse, the region is classified by some as a "physiographic province," an area similar in geologic structure, and by others as a "natural region," an area with similar climate, vegetation, and physical features like elevation (fig. 3).

According to the geologist Charles B. Hunt the basic differences between physiographic provinces are structural, referring to uplift, earthquake faulting, bending and folding of rock layers, volcanism, or combinations of these processes. The variances in these processes and the degree to which they have shaped a given area create physically distinct regions and sections therein. Anyone approaching the Rocky Mountains from the plains can see an abrupt change and would generally agree that two distinct regions meet at this physical boundary.

The transition from the gently tilted rocks of the plains to the more steeply tilted rock layers of the foothills marks the general boundary between the western Great Plains and the adjacent Rocky Mountains to the west. This same kind of distinct visual break can also be seen at the southwestern margin of the Great Plains where that physiographic province meets the faulted rocks of the Basin and Range Province in western Texas and southeasternmost New Mexico and along the Balcones Fault Zone in south central Texas where the southeastern Great Plains border adjoins the western margin of the Coastal Plain Province.

Defining the eastern, northern, and southern boundaries of the Great Plains is much more difficult. Except for the boundary in south central Texas, there is no sharp structural break between the Great Plains and the plains regions continuing to the east, north, and south of it. The Great Plains slopes gently eastward from its western boundary to the plains areas to its east. Sediments and sedimentary rock layers beneath the land surface usually appear to be only gently tilted.

In the face of these uncertainties, where do I draw the boundary of the Great Plains? Because this is a geologic work, I will base my boundary on geologic features if at all possible. Those can change over long geologic periods, but such changes require time spans of a magnitude greater than do those related to culture and dependent upon climate. My geologic boundary for the Great Plains and the boundaries of its sections are shown in figure 4 and can also be seen on figure 5. The boundaries suggested previously by most geologists are very close to those in my figure.

The western boundary begins at the break between the Great Plains and the structurally complex Rocky Mountains, running from Alberta, Canada, south to north central New Mexico. From there the line continues south and then southeast along the break between the Great Plains and the easternmost parts of the Basin and Range Province to just east of the Big Bend area of Texas.

The northern border, in Alberta, runs along the south side of the Athabasca River valley, then turns eastward into western Saskatchewan south of the Christina River. From there I have drawn the line at the topographic break of the Missouri Coteau, a low escarpment extending from western Saskatchewan south across North and South Dakota. In Nebraska, recent geologic events left a sometimes thick mantle of sedimentary deposits that mask where the Missouri Coteau might have been so that the line cannot be extended there with any certainty. For this reason I have drawn the line in Nebraska and northeastern Kansas along the approximate western boundary of the Pleistocene continental glaciers. This line is marked by the end of

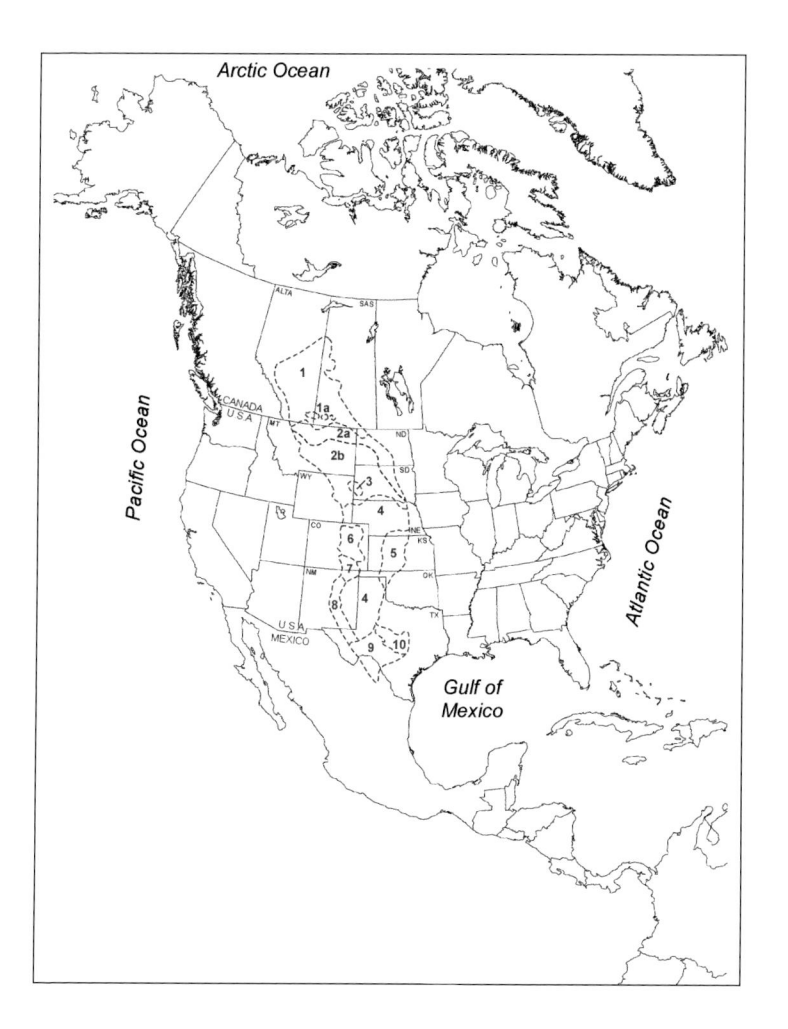

Fig. 4. Author's concept of the boundaries of the Great Plains and its sections. Sections are (1) Alberta Plain (1a is Cypress Hills); (2) Missouri Plateau ([2a] glaciated; [2b] unglaciated, separated by dashed black line); (3) Black Hills; (4) High Plains; (5) Plains Border; (6) Colorado Piedmont; (7) Raton; (8) Pecos Valley; (9) Edwards Plateau; (10) Central Texas Uplift. Boundaries are dashed because they are approximate. Created by author.

Fig. 5. Image of part of North America including the area of the Great Plains. Notice first the Great Lakes. Then look westward to see several features separating the Great Plains from adjacent provinces. The Missouri River valley forms the eastern edge of Nebraska and extends northward through the Dakotas into Montana and the eastern edge of the Rocky Mountains. The Missouri Coteau escarpment extends from southeastern South Dakota northward into Saskatchewan and Alberta. The eastern edge of the High Plains subdivision extends from eastern Nebraska southwestward to Texas. The eastern edge of the Rocky Mountains forms much of the western border of the Great Plains. Used with permission of Paul Morin, University of Minnesota.

glacial till deposits or erratic boulders either at the land surface or found in drill holes.

The line then continues southwest across Kansas, western Oklahoma, and western Texas following the eastern edge of sedimentary rocks of Cretaceous age and then the eastern edge of the Miocene Ogallala Group. I have then drawn the line along the approximate northern boundary of the Central Texas Uplift, then south and west along the Balcones Fault Zone (finally another structural break) to the Rio Grande in the vicinity of Del Rio, Texas.

Most of the earlier maps showing a Great Plains boundary line have stopped the line at the Rio Grande. I have continued the line across the border into a small part of northeastern Mexico, as some Mexican mappers have done. This makes sense to me because the relevant geologic formations and structures do not stop at the river in Texas but continue on into Mexico. This continuation can easily be demonstrated either by reviewing geologic and topographic maps of the area or by standing on the Texas side of the Rio Grande and looking across the river at the rocks exposed along the valley sides to the south in Mexico.

To further complicate the picture, different geologists have divided the Great Plains Province into ten subdivisions called sections (fig. 4). The boundaries between sections are sometimes fairly sharply defined by a structural element like a major fault or fold, but most have boundaries that are more arbitrary. Please see appendix 1 for more information on these subdivisions.

How Did the Great Plains Come to Be?

Now that I have established a boundary for the Great Plains, you might ask how this physiographic province came to be. The major characteristic of the region is a fairly flat land surface. Here and there, however, the land surface changes. How did that happen? What explains the appearance of the Great Plains

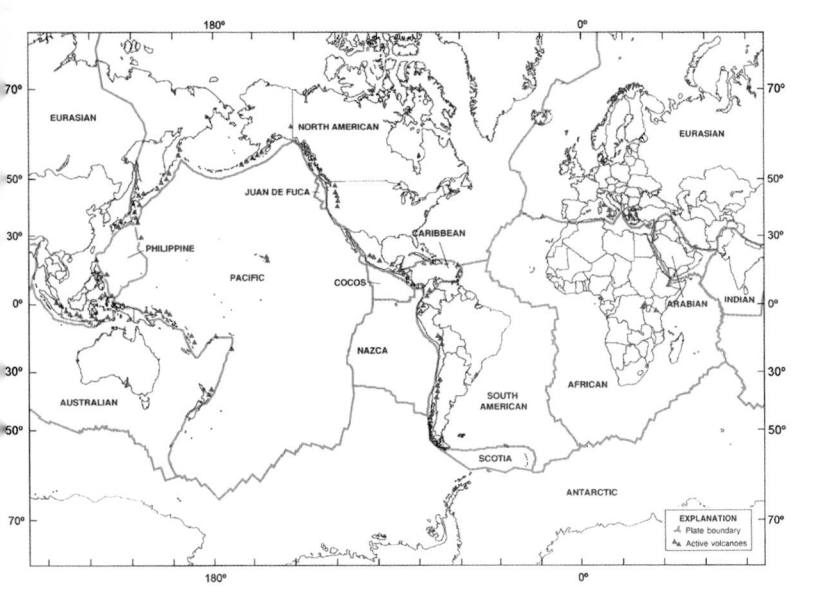

Fig. 6. Current map of principal world tectonic plates, their names, boundaries, and active volcanoes. From Brantley, *Volcanoes of the United States*.

today? The results of geologic change over millions of years provide some answers to these and other questions, such as:

Why does the Great Plains tilt downward toward the east?
Collisions of tectonic plates along the western edge of North America produced folds and earthquake faults (fig. 6). The Rocky Mountains were pushed up along some of these faults. As the mountains rose, so also did the plains to the east, where they rose highest nearest to the mountain front. The result is a land surface that tilts downward to the east (fig. 7). After the faulting occurred, increases in the volume of mineral crystals in the rocks buried deeply beneath the mountains and the nearby plains further uplifted the entire region to the altitudes of today.

Fig. 7. View to the northeast showing the Southern Rocky Mountains Front Range and adjacent Great Plains near Cañon City, Colorado, with the Rockies to the west and the plains sloping to the east. Photo by author.

Fig. 8. Bird's-eye view of the Black Hills, looking north. From Newton and Jenney, *Report of the Geology and Resources*.

Why do the Black Hills stick up in the middle of flat lands?
The rocks beneath the Black Hills of South Dakota were pushed
up by compression forces coming mostly from the west to form
a huge elongated, blister-shaped fold that is generally steeper
on the eastern side (fig. 8). After the folding, the rocks decayed
by exposure to the atmosphere and living organisms and winds
and rivers eroded parts of the uplifted rocks to produce the
rugged, mountainous landscape that we see today.

**Why are there so many volcanic cones, rocks, and hardened
lava flows on the Great Plains in southeastern Colorado and
northeastern New Mexico?** A line of weakness or lineament
in the Earth's crust runs from east central Arizona to the Great
Plains of northeastern New Mexico and southeasternmost
Colorado. From time to time the rocks beneath the surface
here have broken, and molten rock from deep beneath the
Earth has welled out onto the land to form the cones and flows
(fig. 9). Some of these volcanic rocks formed less than 20,000
years ago and appear fresh and new.

**Why is the High Plains Aquifer thickest and most wide-
spread in Nebraska and thinner to the north and south?** The
water in the High Plains aquifer is contained in pore spaces in
river deposits filling many ancient valleys. The rivers in these
valleys drained much of the Rocky Mountains and adjacent
plains. In Nebraska, wide and deep valleys sloped eastward and
northeastward across the state, the general trend of the thickest
and most widespread deposits of the aquifer. Ancient rivers to
the north in the Dakotas and to the south in the Great Plains
of Kansas, southern Colorado, western Oklahoma, western
Texas, and eastern New Mexico did not carve out such deep
valleys and generally did not leave behind deposits as thick as
those in Nebraska.

Fig. 9. View to the south from the top of Mount Capulin in northeastern New Mexico. The two satellite cones and lava flow ridges shown formed in the same volcanic episode that formed Mount Capulin. Photo by author.

Where did the sand in the Nebraska Sand Hills and other sand dune areas of the Great Plains come from and why? Ancient rivers that crossed the plains in the past carried gravel and sand grains and deposited them in their valleys. Dunes in the Nebraska Sand Hills and other parts of the Great Plains were formed when exposed sand, left behind when the courses of the rivers changed, was picked up by winds and deposited in the form of dunes during prolonged major droughts (fig. 10).

What caused the "prairie potholes" of Southern Alberta, Saskatchewan, and eastern North and South Dakota? When the ice sheets covering this part of the Great Plains melted, debris carried by the ice was left behind as a blanket-like deposit covering much of the bedrock. During the melting, pieces of ice sometimes broke off the sheet and were buried in the

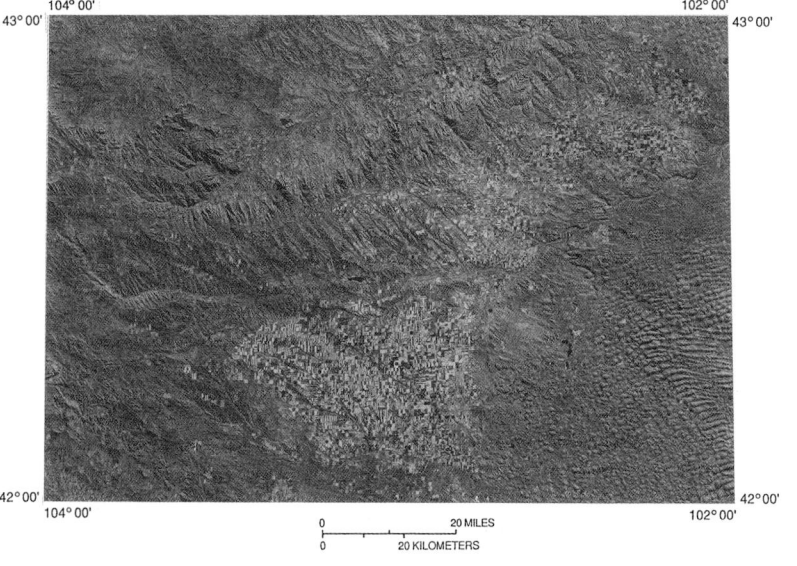

Fig. 10. The Sand Hills are the elongate, generally west-east-trending hills on the right side of this image. The city of Alliance, Nebraska, is at the bottom center; irrigated agricultural fields are in the lower center; and the Pine Ridge is in the upper center and upper left. Synthetic-aperture radar imagery x-band of the Alliance Quadrangle, Nebraska and South Dakota. From *Alliance, Synthetic-Aperture Radar Imagery X-Band.*

glacial deposits. When this buried ice melted, it left behind basin-shaped depressions on the land surface that filled with precipitation or groundwater. These wetland areas are often called "prairie potholes" or "kettles" (fig. 11).

Is there any evidence on the Great Plains related to the end of the Age of Dinosaurs? Most geologists agree that tectonic and climatic effects of large meteorite or comet impacts and major volcanic activity at the end of the Cretaceous Period (see fig. 1) marked the end of the Age of Dinosaurs. At this

Fig. 11. Prairie pothole lakes on the Wisconsinan glacial plain near Wing, North Dakota. Wikimedia Commons, U.S. Fish and Wildlife Service Headquarters.

time the Great Plains area was covered by a sea. The impacts produced tsunamis that left distinctive kinds of sediments. The youngest layers of Late Cretaceous sediments in the Great Plains are enriched with the element iridium and also contain minerals indicating meteorite impacts.

CHAPTER 2

Geologic History of the Great Plains

A brief geological history of the province can help you appreciate the many features that you can see while traveling on the Great Plains. But first you must understand some basic geologic concepts.

Some Geologic Concepts

Today most geologists view the Earth as a single system with the atmosphere, hydrosphere, lithosphere, and biosphere all interconnected to produce the world that we inhabit.

A few geological discoveries and general geologic principles (see appendix 2) will be useful for you to know. In 1669 Nicolas Steno, a Dane, wrote a small book outlining three key principles still used in the study of sediments and sedimentary rocks (fig.12 A-C; bottom layer, labeled 1, is oldest). Sediment, carried into water, moves to the floor of the body of water (anything from a mud puddle to an ocean basin), where it accumulates in nearly horizontal layers (the principle of original horizontality). The layers accumulate one at a time, with the first deposited at the bottom and others deposited above it (principle of superposition). The layers are continuous across the body of water (principle of lateral continuity).

James Hutton, a Scot, published his *Theory of the Earth with Proofs and Illustrations* in 1795. In this and later publications, Hutton and others explained that the Earth's rocks were formed

Air fall
Runoff
Runoff
Lake
Older rock
A
Air fall
Runoff
Runoff
Lake
1
B
Air fall
Runoff
Runoff
Lake
2
1
C
2
2
1
1
3
D

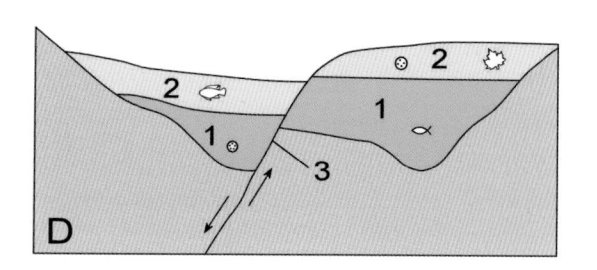

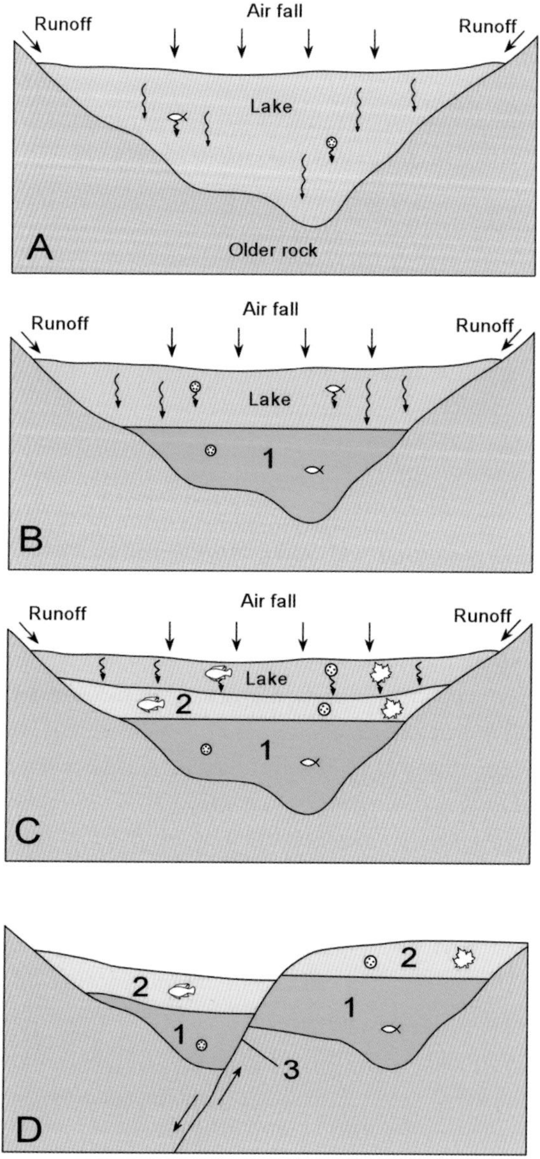

in cycles. Mountains form and partially erode away and sea levels rise and fall repeatedly through geologic time. Hutton observed that any geologic feature, such as a fault, that cuts across another must be the geologically younger feature, an observation that later became known as the law of crosscutting relationships (fig. 12 D).

Then, in 1815, William Smith, an Englishman, prepared the first geologic map of Great Britain. Smith established the principle or law of floral and faunal succession (fig. 12 A–D). That is, fossil species occur in layers deposited in sequence one at a time, with the oldest at the bottom and the youngest at the top. Species succeed one another in a definite and recognizable order, and each geologic formation can have flora and fauna different from the formations above and below. This understanding has allowed geologists to match up layers of rock of roughly equivalent geologic age locally and worldwide and to establish a worldwide geologic rock column and relative timescale.

The development of other fundamental concepts continued. While correlating layers of sediment and sedimentary rock using the fossils they contained, geologists discovered that there was often an abrupt change in fossil types below and above major surfaces of erosion or nondeposition in the rock column. These surfaces, which James Hutton noted in 1795, signal gaps of time between the sediment deposits. These gaps are called unconformities. Geologists have designated the rock sequences between major unconformities as "systems."

Fig. 12. Model cross-sections through time illustrating Steno's principles of superposition, original horizontality, and lateral continuity (12A–C; layer 1 is the oldest lake deposit); the principle of cross-cutting relationships (12D; an earthquake fault cuts through and offsets all layers); and the principle of floral and faunal succession (12A–D). Created by author.

Starting in 1877, Grove Karl Gilbert wrote on the development of river valleys. He noted that, in most cases, rivers have, through time, eroded the valleys in which they flow.

By 1913, scientific technology had reached a point of development where isotopes of elements could be detected and their proportions in minerals and rocks could be measured. In that year, Arthur Holmes of Great Britain used ratios of uranium and lead isotopes in minerals found in some rocks in the geologic column to make the first absolute geologic timescale. Since then, improved technology and new dating techniques have allowed for further refinement of the timescale (fig. 1). Figure 1 is not Holmes's original timescale but the most recent iteration. Note that there are parallel terms in the table for the rocks deposited in certain time spans (erathem, system, series) and for the actual spans of time (era, period, epoch). Geologists separate these so that they know if they are talking about rocks or about the time when they formed.

Events in Earth's History

I pointed out in 1991 that it is impossible to separate the geologic development of the Great Plains from the development of the Earth as a whole. The principal drivers of geological changes on Earth through time are the sun's radiant energy and the Earth's internal heat. After life-forms appeared, they became a third driver. The amount of solar energy that arrives on Earth has generally increased through time over billions of years, but it also varies on human timescales.

The Earth, once fully molten, began to cool about 4.6 billion years ago to its current state. Throughout Earth's history, convection currents have circulated hotter materials from the mantle, the part between the Earth's core and its outer layer or crust, toward the surface and have moved cooler materials back down. As the Earth's crust developed by the concentration of elements with lighter atomic weights at the surface and the oceans and atmosphere developed through the process of outgassing, the ocean basins began to form.

Early in Earth's history, around 3.8 billion years ago or earlier, continents began to form. The minerals and rocks of these lands were subject to the weathering and erosional processes prevailing at those times. Some of these processes may have differed from today's because then the atmosphere contained far less free oxygen and nitrogen and the oceans far less salt.

Also, the oceans may not have been as deep as today because today's deepest ocean basins may be due, in part, to overall cooling and contraction of rocks beneath the sea floor. Continents continued to build up and break up through time as convection currents in the mantle moved oceanic crust and land masses around the surface and recycled oceanic crust.

It is clear that the climate has never been static. In the 1950s, when "continental drift" was being discussed as a theory, geologists studying rocks in the southern hemisphere discovered erosional features and deposits of sediments in Precambrian rocks in Africa, South America, Antarctica, Australia, and India that were formed or deposited by ice sheets. Today, ancient glacial striations (parallel lines or grooves) on those older rocks extend in different directions on these continents. When, however, scientists fitted the continents together like pieces of a jigsaw puzzle, they found that the striations all radiated out from a central area on the reconstructed supercontinent they called Gondwanaland.

The geologic record of marine sedimentary rocks covering parts of the continents and islands makes clear that relative sea level has risen and fallen many times over Earth's long history. These rises and falls were due, in part, to waxing and waning of continental ice sheets, but they were also caused by changes in the shapes and depths of the world's ocean basins due to increases and decreases in the numbers of oceanic ridges and to variation in the rates of spreading and of thermal activity along the ridges.

When I was in grade school, I had a chance to look at the moon through a telescope and saw the craters there. That celestial body was pockmarked with them; in some cases the craters overlapped. Since those days, NASA spacecraft have flown by the other stony planets in our solar system and have sent back images of their cratered surfaces. The only reason Earth does not look like those planets is because life and plate tectonic movements have affected the surface through time, reducing or eliminating the evidence of past impacts.

We now know that impacts have had a profound effect on Earth and on its life at several times in the past (fig. 1). Meteorites striking ocean basins could have led not only to sea floor deepening but to changes in mantle convection and rock melting as well. Particularly at the end of the Paleozoic and Mesozoic eras, enormous numbers of plant and animal species died out due to the direct and indirect results of big collisions.

Most of these collisions did not occur on the developing Great Plains, but known or inferred impact structures in rocks have been found in Texas, Kansas, Iowa, North Dakota, and Montana. A significant impact occurred near the end of the Cretaceous Period when a large meteorite hit the floor of the shallow sea covering the plains area near Manson, Iowa. Even if this and other meteorites did not directly hit the Great Plains, these extraterrestrial bodies influenced the area by producing tsunamis and changes in the atmosphere.

Finally, the fossil record makes clear that life affects the natural cycles of rock weathering, erosion, sediment transport, deposition, plate motion, and climate. The fossils in the oldest rocks are viruses and bacteria. Algae followed them. Fossil cyanobacteria have been found in cherts 3.2 billion years old in western Ontario, Canada.

From that time to the Devonian Period, most sedimentary rocks contain fossils of organisms that lived in the oceans. These organisms changed the chemistry of the oceans and the atmosphere, thereby changing the rates of physical and chemical weathering of rocks exposed on the land. The transport of those weathered rock materials from the land to the ocean basins by winds and rivers increased.

Plants and animals began colonizing land as early as the Silurian Period. Once established, they changed the chemistry of the soil, consequently affecting rates of physical and chemical weathering, climate, and natural carbon sequestration that further transformed the chemistry of all bodies of water on the Earth's surface and of the atmosphere.

The fossil record demonstrates that life has continued to evolve on land and in the seas. During most of that time, organisms used carbon to make wood, calcium carbonate shells, organic materials, and other compounds, at least some of which was buried in sediments. Once humans began using fire and exploiting these buried resources for energy and building materials, more of this carbon was released.

Events in the Great Plains

While they are part of the larger story of Earth's development, the rocks and fossils of the Great Plains also reveal the course of its particular history.

TECTONICS I

All of the present-day continents have one or more areas, called cratons, that are underlain by ancient rocks. These areas are stable and have not extensively changed or deformed for at least the past one billion years (since late in the Precambrian). Cratons include exposed areas of ancient Precambrian rocks called shields, as well as adjacent areas called stable platforms that are covered by a relatively thin veneer of post-Precambrian rocks. The Great Plains overlies a part of the stable platform of the North American craton (fig. 13). We know the general size of the North American craton both because we can see the shield area directly and because geologists have studied the rocks buried beneath the surface of the stable platform area. Geologists have mapped the craton using probes to measure the rocks' electrical and radioactive properties and drills to recover cores and chips of the rocks from the depths where they are buried.

The craton on each continent built up during the Precambrian as small tectonic plates collided and new rock materials were added through time. The rocks and structures in these individual plates differ from one another. Radiometric dating of these rocks has yielded dates from more than three billion to about one billion years in age. The dates from each piece of plate cluster within a certain range. Given the numbers of bits and pieces of tectonic plates that came together as each craton formed throughout the Precambrian, it seems reasonable to infer that there were more oceanic ridges during the Precambrian than there are today. At the end of the Precambrian, most of the rocks of the North American craton were exposed to weathering and erosion, producing an erosion surface, which if later covered by younger sediments or rocks is called an unconformity.

After the cratons formed, the Earth's tectonic plates continued their activity. The eastern part of the North American plate collided with a volcanic archipelago (or island arc) that sat in the ocean to the east at the edge of an adjoining plate, crumpling the continental crustal rocks there. Subsequent collisions with other crustal plates ultimately brought together developing western parts of Europe and northwestern Africa with North America. These collisions produced the Appalachians and a continuing mountain chain crossing what has

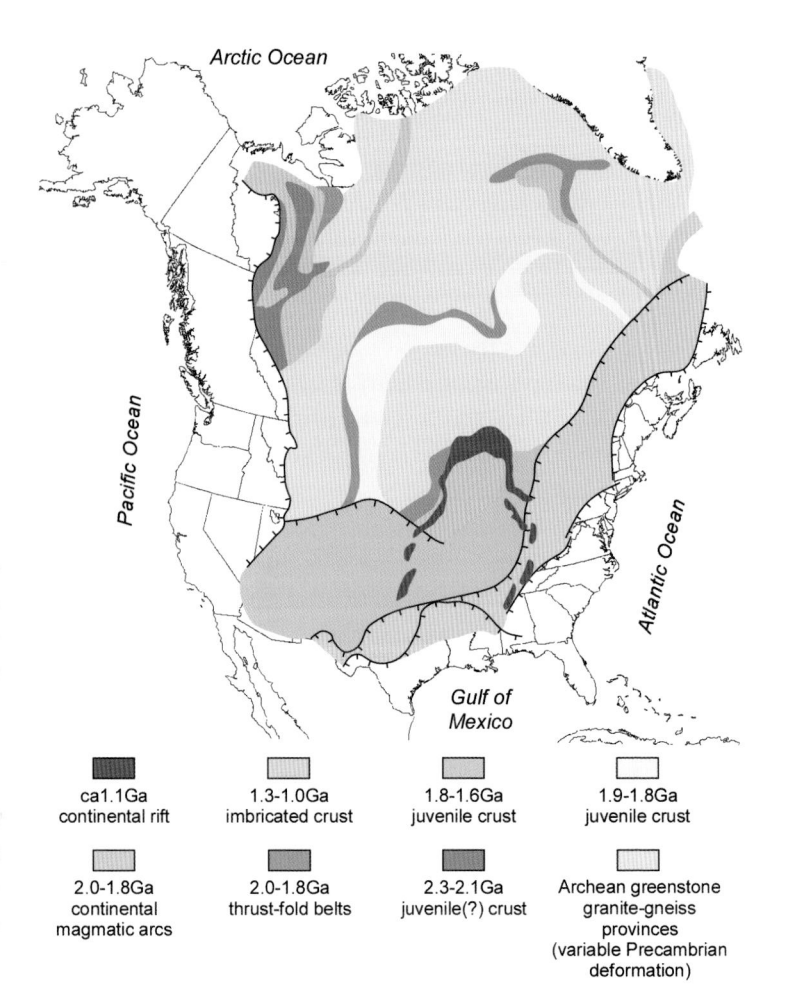

Arctic Ocean

Pacific Ocean

Atlantic Ocean

Gulf of Mexico

ca1.1Ga continental rift	1.3-1.0Ga imbricated crust	1.8-1.6Ga juvenile crust	1.9-1.8Ga juvenile crust
2.0-1.8Ga continental magmatic arcs	2.0-1.8Ga thrust-fold belts	2.3-2.1Ga juvenile(?) crust	Archean greenstone granite-gneiss provinces (variable Precambrian deformation)

Fig. 13. The North American craton, indicated here in colors, was assembled through plate tectonics movements, collisions, and other deformations. "Ga" is billions of years ago. Hachured lines represent faults. Modified, republished with permission of Annual Reviews, Inc., from Hoffman, P. F., United Plates of America, the Birth of a Craton: Early Proterozoic Assembly and Growth of Laurentia, *Annual Review of Earth and Planetary Sciences*, volume 16, 1988, figure 1.

since become Ireland and part of the United Kingdom. The developing western coast of North America was not completely quiet during this interval. A collision in the late Devonian Period deformed the plate margin there.

Across the central craton of developing North America, folding and faulting produced broad domes and basins. These movements formed features like the Illinois Basin, the Ancestral Rocky Mountains (mountains formed near the end of the Paleozoic Era), and the Denver Basin in Colorado. From time to time, volcanoes erupted and left behind ash beds on the floors of the shallow seas covering parts of the craton.

Finally, by the end of the Permian Period, all of the continents and major islands had been brought together by plate motions to form one supercontinent that geologists call Pangaea. Lands and seas never stay put in this history, however. Soon the supercontinent began to break up as plate motions changed directions beginning in the Late Triassic Period. The eastern coast of developing North America was pulled apart from those parts of Western Europe and northwestern Africa to which it had initially been joined. This motion produced rift valleys that today are partially preserved in the Piedmont region of the Appalachian Mountains and in the Connecticut River valley. These valleys filled with terrestrially deposited sediments and volcanic rocks, similar to what is happening today in the East African Rift zone and in the Arabian Sea and Palestine.

Because the relative motion of the North American plate was westward, the western part of North America began and continues to crumple as it collides with the Pacific and Juan de Fuca plates. Bits of continental crust carried on the leading edges of these two plates struck western North America, adding to the North American plate. Today the U.S. northwestern coastal portion of the North American plate is overlapping a descending part of the sea floor beneath the adjacent Pacific and Juan de Fuca plates. At the same time, part of North America just to the south is being pulled apart as it overrides a spreading center coming up the Gulf of California into the western United States (fig. 6).

Pressure from these plate movements on western North America created the Rocky Mountains at the end of the Cretaceous Period. The Black Hills were also formed then. Local deformation and igneous intrusions continuing through the Cenozoic up to relatively recent times formed the smaller mountains and the dikes and sills seen in the Great Plains areas in Montana, southeastern Colorado, and northeastern New Mexico, as well as the intrusions in the northern part of the Black Hills.

To complicate matters further, convection not only drove plate motions through time but also caused the upwelling of molten rocks called mantle plumes or hot spots that remain in fixed positions for longer time spans than do the plates. Eruptions at these hot spots as plates moved across them produced, for example, the islands of Hawaii and also the line of extinct volcanoes across the Snake River Plain in Oregon, Idaho, and northwestern Wyoming, culminating in the Yellowstone caldera. Volcanoes along this line were the sources for the volcanic ash deposits that blanketed the Great Plains during the later Miocene Epoch.

SEA LEVEL CHANGES

Since Precambrian times, relative sea level has risen and then fallen repeatedly. Worldwide changes in sea level can be caused by any number of processes. Sea floor ridges, for example, change in volume and size with changes in tectonic activity. During quieter times, the sea floor deepens as the sea floor plates cool, and the rocks beneath them contract. Overall sea levels

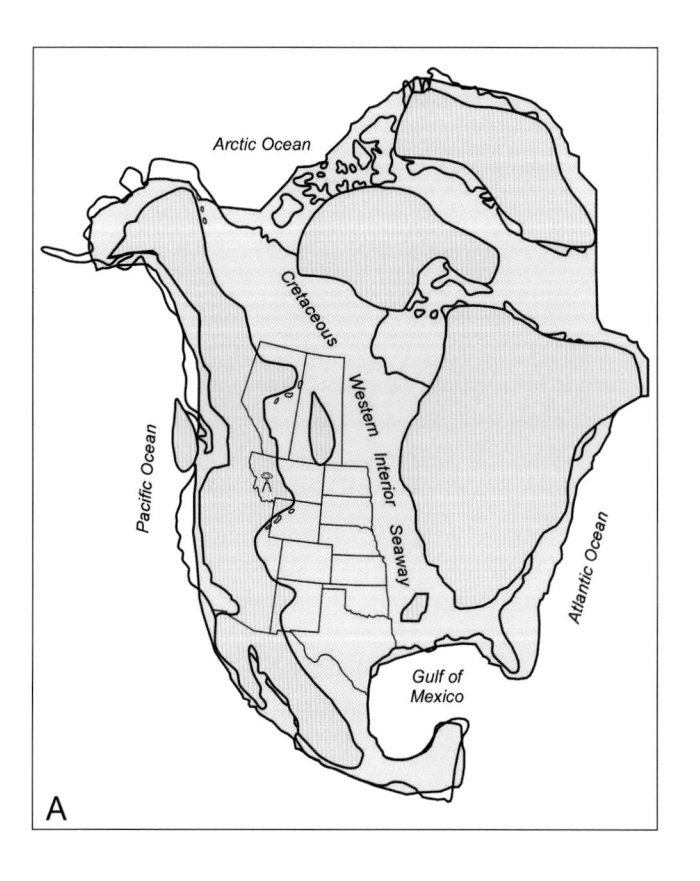

A

Fig. 14. The general area of the Upper Cretaceous Western Interior Seaway during the time of deposition of rocks of (A) the Niobrara Formation and lateral equivalents (83.5 million years ago) and (B) the Pierre Shale Formation and lateral equivalents (72.5 million years ago). The tan color represents land and blue represents the seaway. Notice that the width of the seaway changes from A to B, reflecting changes in sea level through time. Note the symbol for active volcanic areas in western Montana. Modified and simplified based on Gill and Cobban, *Stratigraphy and Geologic History*; Diffendal and Diffendal, *Lewis and Clark*; and Kauffman, "Geological and Biological Overview."

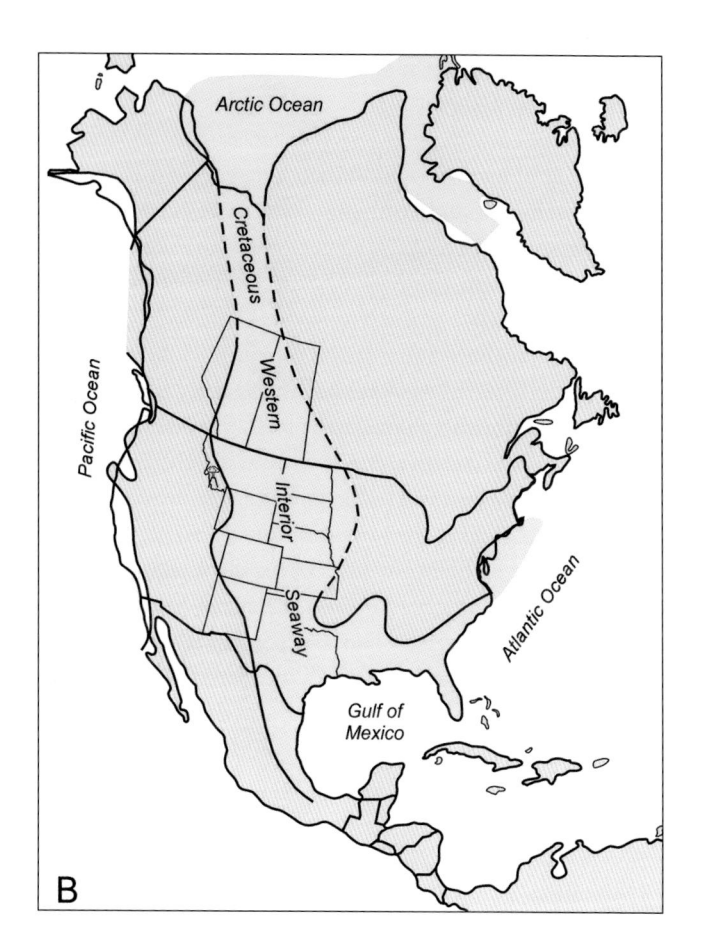

Arctic Ocean

Cretaceous

Western

Interior

Seaway

Pacific Ocean

Atlantic Ocean

Gulf of
Mexico

B

rose from the late Precambrian through much of the Cambrian, leveled off and slightly declined through the Mississippian, dropped considerably and then stabilized until the start of the Cretaceous, and then rose and fell several times, ending in a decline at the end of the Cretaceous. The rise and fall of sea level during the Cretaceous Period has particular relevance to the Great Plains because much of the region's exposed flat-lying bedrock was deposited at this time. (See fig. 14 for a view of what was happening through that period.)

The Cretaceous Western Interior Seaway (fig. 14A) was relatively wide about 83.5 million years ago when the rocks that make up the Niobrara Formation were being deposited. At this time active volcanoes in western Montana shed volcanic ash into the seaway. These ash deposits are preserved in parts of the rock sequence of the Niobrara Formation.

The Niobrara Formation itself is primarily a limestone rock unit made up of microscopic calcareous materials, such as shells, secreted by tiny marine animals and plants. During this high sea level period, lands on either side of the seaway were distant from the center area. Land-derived sediments were deposited near the shore and became less important in the overall sediment composition toward the center of the seaway, where calcareous organic debris accumulated in proportionately greater quantities. Proximity to the shoreline is important to keep in mind when interpreting what went on in the seaway through time.

During deposition of the rocks of the overlying Pierre Shale Formation, about 72.5 million years ago, the Western Interior Seaway was narrower, and the shoreline was closer to the center of the seaway (fig. 14B). Land-derived muds dominated deposition across much of the seaway, which, after compaction, led to formation of the Pierre Shale. Volcanoes were still present and active in part of western Montana, still shedding volcanic ash eastward into the waters of the seaway. These ash deposits, subsequently altered to types of clays called bentonites, are also present in the Pierre.

As sediments were deposited on the sea floor, the adjacent coastal plains, and places farther inland, the weight of these additions together with movement in the crust caused the Earth beneath the coastal plain and the seaway floor to sink through time. By about 66 million years ago, the Rocky Mountains were forming and the sea was withdrawing from the midcontinent for the last time. The sea's retreat left behind mostly beach (Fox Hills Sandstone) and coastal plain (Hell Creek Formation) deposits overlying the Pierre Shale, which are similar to those deposited today on the land bordering the Gulf of Mexico.

TECTONICS II

Continental deposits from the Paleogene and Neogene Periods of the Cenozoic Era are the second-most-common surface deposits of bedrock in the Great Plains today, the older beds of the Cretaceous Western Interior Seaway being the most common. These Cenozoic rocks units were once more widespread. Today their distribution is more restricted than at the end of the Miocene because rivers have since eroded parts of them. The deposition of these continental deposits is directly related to the Cenozoic tectonic history of western North America.

Plate motions and their attendant volcanism, earthquake faulting, rock crumpling, and warping did not stop at the end of the Cretaceous. The greatest number of volcanic impacts to the northern Great Plains, as recorded by formations composed of ash debris, occurred during the Cenozoic from 37 to 17 million years ago. Later explosive eruptions were still large, but less frequent. In its westward movement, the North American plate passed over the Yellowstone hotspot and collided with the Pacific and Juan de Fuca plates (fig. 6) to cause major volcanic eruptions in the Rocky Mountains and lands to the west. Many of the volcanoes were explosive and emitted huge quantities of ash into the atmosphere. From western Nebraska northward to western North Dakota, thick rock formations composed mostly of airfall ash debris blanket extensive areas of land covering the Cretaceous seaway deposits. In western and central Nebraska, these volcanic airfalls are more than 1,000 feet thick in places.

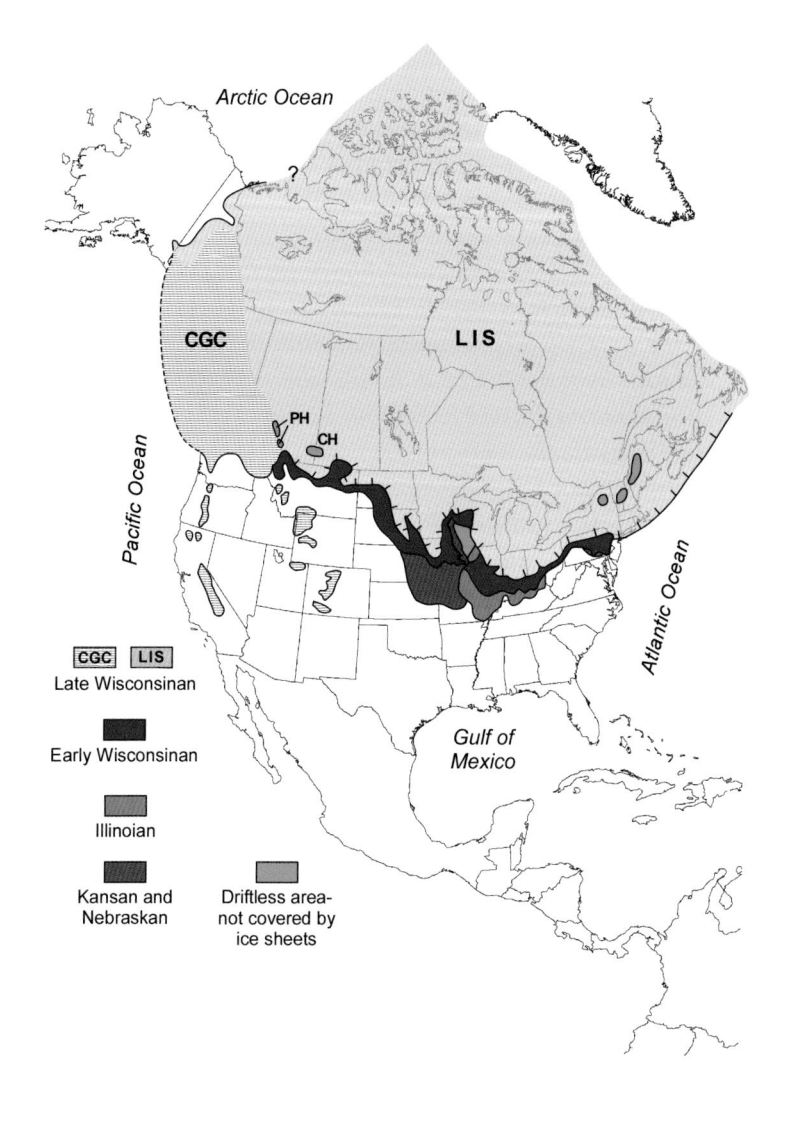

Renewed uplift in parts of the Rocky Mountains, particularly from the middle Miocene into the Pliocene, caused erosion of the mountains by rivers. Huge quantities of sediment were deposited by these rivers in valleys on the Great Plains from South Dakota southward to western Texas and eastern New Mexico. Lenses of relatively pure volcanic ash in this pile of mostly river deposits are evidence of continued volcanic activity during the Neogene Period.

ICE AGES

Worldwide waxing and waning of glaciers due to cycles of climate change also caused sea level changes many times during Earth's history (fig. 1). The most recent Pleistocene glaciations, which began about 2.588 million years ago, had a significant influence on the development of the Great Plains as we know it, even in places away from the edge of the ice sheets. Figure 15 shows the traditional view of the distribution of the North American Laurentide Ice Sheet (LIS) and mountain glaciers through the Pleistocene. The names of major glacial advances are in flux, but the oldest ones, formerly called Nebraskan and Kansan, left behind deposits when they melted away. The more recent Illinoian glaciation extended the farthest south of all of them but did not extend as far west as the earlier ones.

Fig. 15. Generalized map of the areas covered by glacial ice during the Pleistocene. LIS is the Laurentide Ice Sheet, CGC is the Cordilleran Glacial Complex and equivalents in the western United States, PH is Porcupine Hills, CH is Cypress Hills. The hachured line marks the maximum former extent of the last Wisconsinan ice advance. Modified and simplified from Gerlach, *National Atlas*; and interpretation of writings in Douglas, *Geology and Economic Minerals*. Traditional Nebraskan-Wisconsinan names have been used here, but note that treatment of these names is in flux; see Roy et al., "Glacial Stratigraphy and Paleomagnetism," for a modern update.

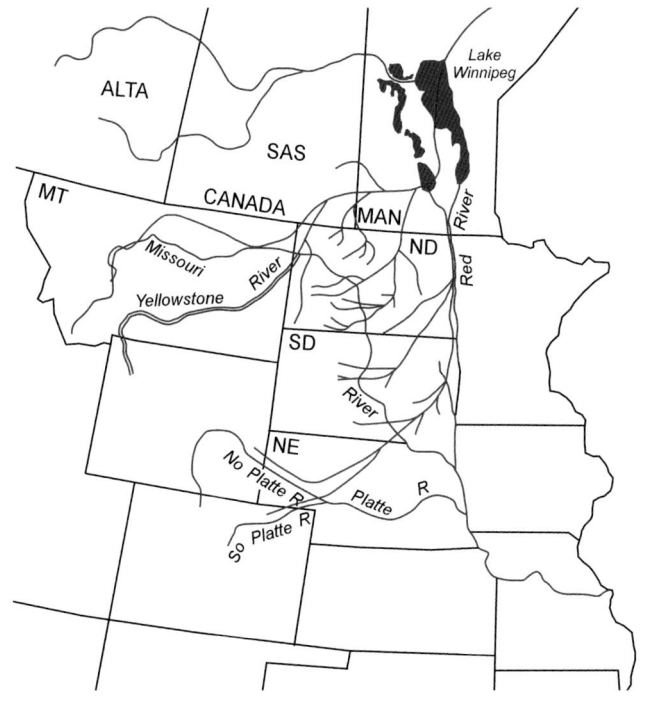

Fig. 16. Probable drainage patterns of the Missouri River and some of its tributaries, including the Platte River System, during the Late Pliocene before the start of the oldest Pleistocene ice age (red) and the present course of the Missouri and some of its tributaries (blue). Notice that the present course of the Missouri River follows the approximate western edge of the LIS Late Wisconsinan ice sheet shown in figure 15. Slightly modified from Diffendal and Diffendal, *Lewis and Clark*.

Throughout the Pleistocene there were so-called driftless areas, a few high spots in the United States and Canada that were never covered by glacial ice.

As ice sheets developed, lands, drainages, and life-forms changed. Vast amounts of water were tied up in the ice sheets, at times lowering sea level worldwide by at least as much as 300 feet. When the ice melted away, vast quantities of water were released, raising sea level and modifying the lands beyond the ice front. Huge masses of sediment carried by the ice were left behind as the ice sheets and mountain glaciers melted away. River valleys, present before the lands were glaciated, were eliminated, and new ones developed as each ice sheet waxed and waned. The Missouri River and its tributaries, for example, were shifted fundamentally from their original courses as a result of the Pleistocene glaciations (fig. 16). In the Late Pliocene the Missouri flowed north into Hudson Bay. The modern system is a result of river valley development during the Late Wisconsinan in the area bordering the western margin of that waning ice sheet.

SUMMARY

The geologic development of the Great Plains began with the formation of the ancient craton, a relatively stable part of the Earth's crust, by crustal movement and collisions of tectonic plates and the subsequent burial of the craton beneath layers of coastal plain and marine sediments deposited in repeated sequences. These sequences reflect rising and falling sea levels and regional uplift or subsidence of parts of the stable shelf rocks from time to time due to faulting and other tectonic plate motions. Marine deposition continued until the end of the Cretaceous and the early part of the Paleogene Periods.

Starting at least from near the beginning of the Cretaceous Period, volcanoes periodically erupted farther to the west of the Great Plains, ejecting large masses of ash into the atmosphere that fell onto the waters of the Western Interior Seaway and accumulated on the sea floor. Volcanism continued in the mountain west throughout the Cenozoic Era, yielding ashfalls on the Great Plains. These falls produced thick blankets of ash during the Paleogene as well as marker layers in younger Neogene Period formations otherwise deposited mostly by rivers.

The major mountain-building episode that produced the Rocky Mountains of the United States and Canada, as well as the Black Hills, began at the end of the Cretaceous. More local deformations and igneous intrusions during the Cenozoic formed the smaller mountains and the dikes and sills on the Great Plains in Montana, Colorado, and New Mexico.

Since the end of the Cretaceous Period, worldwide sea levels have generally declined through the Cenozoic Era, exposing the rocks of the Great Plains to erosional processes. Sediments were removed, principally by river erosion of the mountains, and transported onto the Great Plains, where some of these sediments remain today. Others were eroded yet again and transported generally farther to the north, east, and south.

Over the last 10 million years, regional uplift of the whole Great Plains and Rocky Mountain area has caused major changes in regional climate, the position of river systems, and the distribution of biotas. The several major mountain and continental glaciations, starting at the beginning of the Pleistocene Epoch, further changed the position of rivers, the distribution of biotas, and the deposition of sediments across the Great Plains.

The saga continues today.

CHAPTER 3

Visiting the Great Plains

Each state and province in the Great Plains has many wonderful places that are worth visiting, studying, and enjoying. Some of these have been identified for several of the states in the map *The Top 50 Ecotourism Sites in the Great Plains*. A few are included in this chapter, but most described here are not on that map. I have chosen only ones that I have spent some time exploring and studying.

When driving to visit sites, my wife and I usually take copies of the latest state or provincial highway maps, topographic maps of the areas we plan to visit, geologic highway maps (if they are available), and reading materials on geology and other features. Information about all of the sites detailed below can be found on the Internet. In appendix 3, I have included a list of cautions that you should keep in mind any time that you travel on the Great Plains.

Topographic and geologic maps, including geologic highway maps, are available from state and provincial geological surveys, the U.S. Geological Survey (USGS), and private organizations, as well as in state gazetteers and online. Looking carefully at topographic maps and the shape of the land depicted by the contour lines (lines of equal altitude) will allow you to get a better feel for some of the conditions that you may encounter on your trip. Using them with the geologic maps will allow you

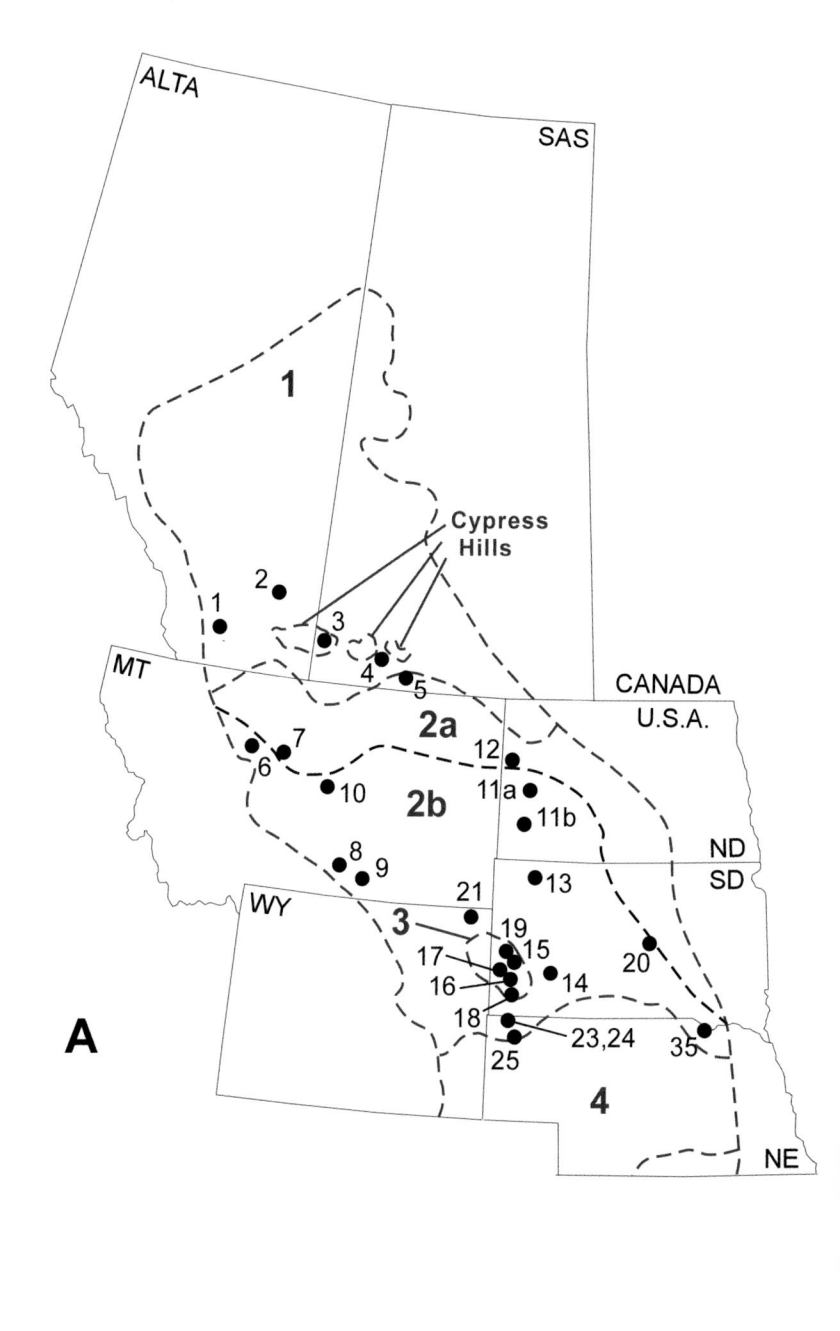

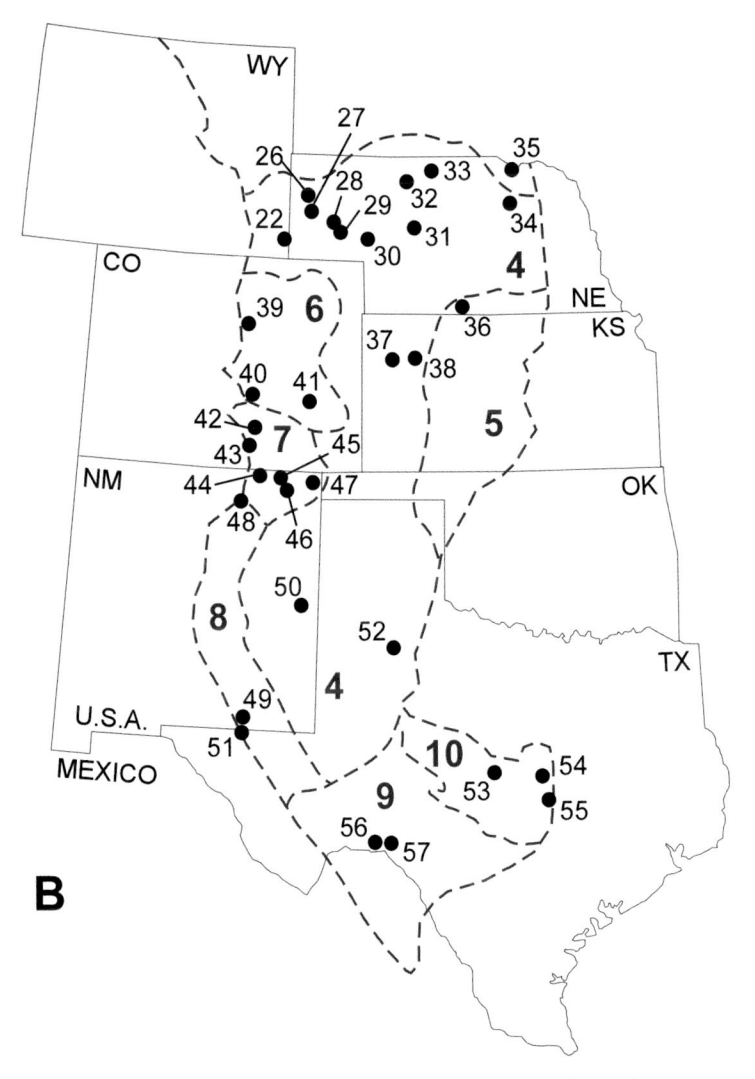

Fig. 17A–B. Maps showing general locations of sites in the northern half (17A) and southern half (17B) of the Great Plains that are described in this book. Black numerals indicate sites. Red numerals refer to physiographic sections of the Great Plains Province (see fig. 4). Created by author.

to match changes in the slope of the land with the occurrences of new formations or of folded or faulted rocks.

The topographic maps will also reinforce what you observe while driving along, specifically that much of the Great Plains is rather flat and that most of the land in both the United States and Canada is crossed by north-south and east-west roads. This land division into more or less square-shaped parcels in a grid is due to both the U.S. and Canadian governments adopting a uniform land survey technique. The U.S. Land Ordinance of 1785 established the method for surveying the western lands. This method was applied to Florida and to all of the states west of the Appalachians except Kentucky, Tennessee, and Texas. A similar system is used in parts but not all of Canada.

This system makes navigation across much of the Great Plains rather easy, so long as you keep the compass directions straight in your mind and you can read a map. Oh, I know that many of you are saying to yourselves that you have a GPS in your car and don't need maps. Perhaps so, but then again you may need those maps if your system fails, something that may happen when you least expect it. In fact, even when in service, GPS is often not totally reliable. The website for Wind Cave National Park, located in the Black Hills of South Dakota, warns, "Do not trust your GPS to find the Visitor Center."

The 57 sites included in this book (fig. 17) are important for their geological, paleontological, or archaeological features. I have used the letters *g*, *p*, and *a* to indicate, following each site name, feature(s) for which the site is best known. Those sites that are on the *Top 50 Ecotourism Sites* map are designated with an *e*.

Throughout this narrative, I will use the names of rock formations and groups of formations. A formation is a three-dimensional mass of rock with one or more features that can be used to distinguish it from formations above and below it. A formation is usually named after the geographic place where it was first described. I will describe formations and groups briefly because of space limitations of this book. I will describe the formations either from the top downward or from the bottom upward, depending on where I began my observations when I visited the site (top or bottom of a hill, for example). Also, because of space limitations, I will not give exact directions on how to reach a specific site unless those are important to the narrative. Directions to sites can easily be found on the web. Almost all of the photographs in figures 18–76 are my own, with the exceptions noted in the captions. I will begin with Canadian sites and work generally south across the Great Plains.

SITES IN
CANADA

FEATURES

G geological

P paleontological

A archaeological

E ecotourism

Head-Smashed-In Buffalo Jump, Alberta

G, P, A

This UNESCO World Heritage Site is located in Alberta, west of Fort Macleod, toward the southeastern end of a geomorphic feature called the Porcupine Hills. It is underlain by parts of the Porcupine Hills Formation, a Paleocene continental deposit of nearly horizontally layered sandstones, siltstones, and mudstones. The sandstones are fairly well cemented and form ledges, some of which are thick enough to appear as cliffs.

Fig. 18. The escarpment at Head-Smashed-In Buffalo Jump, Alberta, and the slope to the lower glaciated part of the Great Plains along the western margin of the Laurentide Ice Sheet. Bison were driven from the area on the left to the right. Photo by author.

You can see where Native Americans herded bison and directed them to the cliff edge, where they tumbled to the base of the cliff about 65 feet below. The site was used intermittently for this purpose starting more than 5,700 years ago. Today the cliff is only about 33 feet high because of subsequent deposition of weathered and eroded rock debris from above. The park visitor center has a large replica of a part of the cliff, with bison at the top and an archaeological dig at the base.

The cliff is not continuous but occurs at the surface now and again along the length of the Porcupine Hills. It looks as if hunters could use any of several other locations for the same purpose, and in some cases, at least, they did. The land below the base of the hills has the topography and drainage typical of the land once covered by ice sheets during the ice ages, with large blocks and boulders scattered on the land surface, having been left behind when the ice sheets melted. The upper parts of the hills have no glacial deposits and were never covered by ice sheets. However, on the western side of the Porcupine Hills the land between the base of the hills and the Canadian Rocky Mountains to the west shows signs of glaciation similar to those on the eastern side. The Porcupine Hills were a high spot between the western edge of the major Laurentide Ice Sheet that covered most of Canada during the last ice age (Last Glacial Maximum) and the Cordilleran Glacial Complex of mountain glaciers and the glaciers that formed from their coalescence on the lowlands adjacent to the mountains (fig. 15).

Dinosaur Provincial Park, Alberta

G, P

This park and the nearby Royal Tyrrell Museum in Drumheller, Alberta, are paradise for dinosaur lovers of all ages. The park is located in part of the valley of the Red Deer River, where sandstones and mudstones were deposited as soft sediments on a large coastal plain from about 76.7 to 74.9 million years ago. During this time, the edge of the Western Interior Seaway covering central North America from the Gulf of Mexico to

Fig. 19. Dinosaur Provincial Park badlands, Alberta. Cretaceous sandstone is in the foreground, sandstones and mudstones in the background. Strata were deposited on the coastal plain bordering the Cretaceous Western Interior Seaway. Photo by author.

the Arctic Ocean was not too far to the east (fig. 14). A number of different names are used for the exposed Cretaceous formations at the site. Some researchers have written that the rocks are all parts of the Oldman Formation, named from the exposures along the Oldman River valley southwest of the park. The Oldman Formation is said to be the upper part of the Belly River Group of formations. Other geologists have noted that there is a widespread erosion surface in the rock sequence and have divided the rocks in the park and areas adjacent to it into an Oldman Formation and an overlying Dinosaur Park Formation of either the Judith River Group, a Montana name, or the Belly River Group. By now your eyes have probably glazed over, but don't be deterred. The Alberta formations can be traced into Montana and are all part of the same coastal plain deposits regardless of the names we give them. For us, Dinosaur Park and Oldman Formations will do for names here.

In the Late Cretaceous, an upland area to the west of the park area shed sediments into rivers draining east and southeast across a coastal plain into the Bearpaw Sea, part of the Western Interior Seaway. The climate at the time has been inferred from the nature of the sediments and their included fossils. The fossils reported so far from the Dinosaur Park Formation include many species of plants (algae, fungi, bryophytes, lycopods, ferns and related species, ginkgos, gymnosperms, and angiosperms), invertebrates (freshwater bivalves and snails), fishes, amphibians (salamanders, frogs), primitive mammals (multituberculates, marsupials, and placentals), and many reptiles (turtles, crocodilians, lizards, plesiosaurs, pterosaurs), including the dinosaurs for which the area is justly famous.

Taken as a group, these fossils indicate that the coastal plain was warm and humid during the deposition of both formations. The sandstones were deposited in channels of rivers and estuaries, the more extensive sandstone and mudstone blankets were deposited on floodplains adjacent to the river, and the ancient soils (paleosols) formed on exposed surfaces at times when sediment deposition was interrupted.

Today the park is partly grasslands, partly badlands (highly eroded areas of bare rock, mostly fine-grained mudstone and claystone, that form when vegetative cover has been destroyed) where the Cretaceous rocks are exposed, and, of course, the Red Deer River. Nothing today looks much like the Cretaceous landscape or most of the Cretaceous flora and fauna. But the truly outstanding fossils from here, exhibited where they were found in the park, in the visitor center, and in the Royal Tyrrell and other museums and scientific collections around the world, all testify to the nature of the former Cretaceous environment of the park. These specimens include many skeletons of at least 35 different species of herbivorous and carnivorous dinosaurs.

Cypress Hills Interprovincial Park, Alberta and Saskatchewan

G, P

The highest point between Labrador and the Canadian Rockies is at the western end of the Cypress Hills in Alberta. The hills are divided into three sections, a West Block, a Centre Block, and an East Block, the last generally unnamed on tourist maps. These areas are all parts of erosional remnants of land that have

Fig. 20. Conglomerate cliffs site on the east end of the West Block of Cypress Hills Interprovincial Park, Saskatchewan. The lower hills and plains beyond the trees were glaciated. The conglomerate is composed of cemented river channel deposits, marking the lowest place on an ancient landscape. Photo by author.

not been worn down by the actions of rivers and ice sheets that scoured down the adjacent plains. There is no evidence of glacial erosion on the higher parts of the West Block.

The West Block has the greatest relief and the best views. The altitude difference between its top and its base is about 2,000 feet, an impressive difference in this area of the Great Plains. You will find outstanding views of the glaciated plains on the many overlooks all around the top of the block.

The slopes are covered by forests that do not occur on the surrounding plains, most of which is today used for production agriculture and grazing. On the approach from the north, the upper flanks and parts of the tops of the hills are covered by forests of lodgepole pine, white spruce, and aspen. Much of the top of the West Block is covered by fescue-grass prairie. Mixed-grass prairie covers most of the eastern parts and dry south-facing slopes.

The highest and most scenic parts of the West and Centre Blocks of the Cypress Hills have been set aside as an interprovincial park. The part of the park in the West Block, which extends across the Alberta-Saskatchewan border, is the most interesting to me because you can see perhaps the best rock exposures and other geologic features there. At the bottom of the bluffs, the formations are largely Late Cretaceous in age. The oldest is the Bearpaw Shale, a black shale deposited in the Bearpaw Seaway, noted in the preceding site description. Above this shale is a marginal marine formation overlain by formations of sandstones, siltstones, and clays deposited in coastal plain settings like the beds at Dinosaur Park. The coastal plain environment and deposits continued into the Paleocene here, but the formation includes some coal beds. An unconformity separates these beds from the overlying gravels, sandstones, and conglomerates of the Cypress Hills Formation (Eocene-Miocene), which cap the tops of the blocks and can be best seen at the Conglomerate Cliffs overlook at the northeastern side of the West Block in Saskatchewan. This conglomerate formation is a valley filling with cross-bedding, indicating that the braided river that brought in the sands and gravels came from the southwest.

The source areas from which gravels come can be determined generally by comparing and matching the types of rock in the gravels with the rock types in potential mountain source areas, a technique that geologists have used at least since the late 1800s. The gravel clasts here are mostly quartzite, with lesser amounts of other igneous and metamorphic rock types from either the older cores of local uplifts on the plains, like the Sweetgrass Hills, Bearpaw Mountains, and Highwood Mountains, or from the Rocky Mountains in Montana.

Skeletal elements, including those from fossil amphibians, snakes, crocodilians, turtles, brontotheres, rhinos, huge pig-like enteledonts, rodents, insectivores, camels, saber-toothed cats, and deer-like animals have been found in the formation, mostly in rocks in the East Block.

The question of why the Cypress Hills were not eroded away by past and present rivers draining the area and by the Pleistocene ice sheets is an open one. Some have proposed a structural cause, because the formations have an eastward plunge with some associated faulting. Both the base and the top of the Cypress Hills Formation slope eastward, and this slope may be due to folding. On the other hand, the geometry of the Cypress Hills Formation resembles parts of other Cenozoic river deposits across the Great Plains, and the slope may simply be the original slope of the deposits. There are many such examples of hills capped by conglomerates and gravels across parts of the Great Plains. These were left behind on high lands when rivers shifted and down cut through less permeable formations to make new valleys, thus producing an inverted topography where older river deposits cap hills (see also site 39).

The Cypress Hills are still eroding away. Stream runoff erodes the rocks from the sides of the hills, and landslides affect many slopes. Horseshoe Canyon and Police Point in the West Block are two examples.

T. rex Discovery Centre, Saskatchewan

G, P

Driving into the Frenchman River valley and approaching the village of Eastend, Saskatchewan, the light pastel shades of the mudstones and sandstones of the Upper Cretaceous formations shout out "continental deposits" and "DINOSAURS." Dinosaur bones have been found here for many years, but in 1991 a bone from a *Tyrannosaurus rex* skeleton was discovered by a local school principal. In 1994 paleontologists found and excavated much of a *T. rex* skeleton, later nicknamed "Scotty."

Fig. 21. A *T. rex* skull in the *T. rex* Discovery Centre, Saskatchewan. Photo by author.

Subsequently, a research and education center was built to house the specimen and other finds. In May 2000 a new, large museum building built into the side of the valley north of the village and the river, dubbed the *T. rex* Discovery Centre, opened for visitors. The center subsequently became part of the Royal Saskatchewan Museum (RSM). The center building is primarily a museum with excellent displays of dinosaurs and other fossils, but it has a theater and a paleontology preparation lab among the other facilities. Kids of all ages who are interested in dinosaurs will enjoy this wonderful place.

Grasslands National Park, Saskatchewan

G, P

This beautiful national park in south central Saskatchewan is divided into two blocks, West and East. The entranceways to the two are separated from one another by about 100 miles of mostly paved roads. The park headquarters building, with some geological displays, is in the village of Val Marie on the west side of the West Block. This block, located in the valley of the Frenchman River, has the easier driving access of the two.

Fig. 22. Cretaceous marine and coastal plain deposits at Grasslands National Park, West Unit, 70 Mile Butte Access, Saskatchewan. Photo by author.

There are several choices of places to drive along and to walk within the West Block. One is to take the West Block Ecotour Scenic Drive from Val Marie east and then south into the park. The land surface is covered by alluvium in valleys and glacial tills on the uplands. Glacial boulders or erratics have been left in piles along the edges of farmland and in pastures. Once you cross the park boundary and approach the Frenchman River, Upper Cretaceous mudstones and sandstones crop out along the valley sides and tributaries to the river. I also recommend a second route, the 70 Mile Butte Access south and east of Val Marie. At 70 Mile Butte, walking trails cross the Cretaceous badlands. The trails lead walkers past springs, landslides, badlands, and interesting flora and fauna, which may include one of my favorite animals, the prairie rattlesnake. I was unaware this species ranged this far north until we started planning our trip to the park.

The eastern side of the East Block is more interesting geologically, though more difficult to access. The Upper Cretaceous Bearpaw, Eastend, Whitemud, and Frenchman Formations and the Paleocene Ravenscrag Formation, discussed earlier, crop out from the floor of the valley of Rock Creek through the Rock Creek Badlands to Zahursky Point. Marine fossils occur in the Bearpaw Formation and dinosaur fossils in the overlying Cretaceous formations. Glass beads and iridium-enriched clays from fallout from the Chicxulub Impact in the Yucatan mark the Cretaceous-Paleogene boundary and the extinctions of the dinosaurs and many other organisms. If you haven't observed that boundary, this is one place to see it! Another is Sugarite Canyon State Park, New Mexico (site 45).

SITES
IN THE
UNITED
STATES

MONTANA

Giant Springs State Park

G

I've seen many large springs, but Giant Springs is exceptional even though it flows into a reservoir behind a dam on the Missouri River at Great Falls, Montana. The discharge is approximately 338 million gallons per day. The runoff from the springs into the south side of the river must have been impressive when Meriwether Lewis and William Clark and their Corps of Discovery passed by on June 18, 1805. Lewis remarked that he thought this was the largest spring then known in America and that its waters kept their clear, blue color for

Fig. 23. Giant Springs, Great Falls, Montana. Part of the impounded Missouri River shows in the background. The springs are fed by artesian water flowing through underground cavern systems in Mississippian limestone formations. Photo by author.

about one-half mile downstream to where they entered the turbid Missouri.

The springwaters emerge from the deeply buried and cavernous Madison Limestone of Mississippian age at the surface along fractures in the overlying Lower Cretaceous Kootenai Formation. The Madison aquifer is recharged at exposures in the Little Belt and Rocky Mountains tens of miles away and topographically much higher than the springs. The beds of the Madison dip downward from the recharge area, putting the water under considerable hydrostatic head by the time it reaches the area beneath the springs.

Upper Missouri Breaks National Monument

G, P, E

Part of the Upper Missouri National Wild and Scenic River flows through this monument in Montana. Geologically, the site clearly illustrates Steno's principles (fig. 12). The rock layers are horizontal and the oldest is at the bottom. Younger masses of black rock cut across the other layers. The only problem is that much of the area is inaccessible by car. You can, however, see parts from highways and gravel roads and much if you

Fig. 24. Upper Missouri Breaks area from the south side of the Missouri River in Montana. The black rock layers are dikes and sills of hardened magma; the white rock in the foreground is Cretaceous sandstone. Dikes illustrate the principle of cross-cutting relationships. Photo by author.

take one of the river tours along the Missouri departing from the town of Fort Benton. Lewis wrote about the area in great detail and made many accurate geologic observations here in late May 1805, captivated by the great beauty of the scenery and the geology.

Bedrock exposed in the park is mostly sandstones and shales containing fossil plants and animals that lived on land or in shallow ocean waters near shore. Most of the bedrock is of Upper Cretaceous formations deposited in coastal plain and offshore settings in the Western Interior Seaway. From oldest to youngest, these are the Eagle (including the white-colored Virgelle Sandstone), Claggett, Judith River, and Bearpaw (our old friend from some Canadian sites) Formations. Cenozoic river deposits cap the uplands north of the Missouri River in the eastern part of the monument and glacial deposits occur on some of the higher ground mostly north of the river. The Cretaceous coastal plain deposits contain fossils of land plants, and the marine Bearpaw Shale yields fossils of ocean (marine) animals.

Structural and igneous intrusive features affect the rocks in the monument. You can view bending and offsets of the rock layers due to faulting and folding of the rock along faults on the valley sides of the Judith River near its confluence with the Missouri in the south central part of the monument. Dikes and sills of black, hardened magma of Eocene age are visible crossing the valley sides of the Missouri. Because they are harder and more resistant to erosion than the sedimentary rocks they intrude, they stand up like walls crossing the landscape. The dike rocks are fractured in most places, giving the fractured rock a honeycomb shape. Similar intrusive structures are more easily accessible at Spanish Peaks (site 43) in Colorado.

Pompeys Pillar National Monument

G, P, A

This butte on the south bank of the Yellowstone River in Montana is composed of weakly cemented sandstone that is part of the Hell Creek Formation (Upper Cretaceous), a marginal marine deposit. The sandstone crops out on the north riverbank as well. The remnant is best known because William Clark stopped here as he and some members of the Corps of

Fig. 25. Pompeys Pillar, Montana. This erosional remnant of sandstone was deposited on a coastal plain during the Late Cretaceous. Photo by author.

Discovery traveled down the Yellowstone to meet the rest of the party at the Yellowstone-Missouri confluence. Clark carved his name and the date of the visit into the rock on July 25, 1806. That graffiti has endured until today and is the only surviving evidence of the exact place where a member or members of the expedition stopped during their journey across the Great Plains.

Fossils in the sandstone, along with the nature of the rock and its history, are interesting to contemplate. Presently, the floodplain of the river extends south of the rock and of the present riverbank. This suggests that, in the past, the river flowed south of the "pillar" and subsequently changed its course by cutting through the Hell Creek sandstone. Rivers are strange and wondrous in their ways.

Little Bighorn Battlefield National Monument

G, A

Visit this Montana monument to better understand history by observing the actual terrain on which the battles took place. Look also at the landscape and the geology. The site is on the eastern side of the Little Bighorn Valley between the river and some tributary drainages. During the battles, George Armstrong Custer and his men were on high ground. The weathered rock beneath the soil at the battlefield is from decomposition of Upper Cretaceous Western Interior Seaway coastal plain and marine deposits (Judith River and overlying Bearpaw Shale Formations). Potable water is at a premium here, and shade is sparse.

Fig. 26. Little Bighorn Battlefield National Monument, Montana. Scarce water makes this hillside indefensible over the long term. From Wikimedia Commons, by 1025wil.

Judith Mountains

The Judith Mountains are one of the several isolated mountain ranges in Montana, including the Sweetgrass Hills, Bearpaw Mountains, Little Rocky Mountains, Highwood Mountains, and Moccasin Mountains. Meriwether Lewis noted them in his journal and correctly concluded that they were not the start of the Rocky Mountains because they were not continuous mountain ranges.

Fig. 27. The Judith Mountains, east side, with the adjacent plains of Montana in the foreground. When these areally small mountains uplifted, they deformed older strata. Photo by author.

The Judith Mountains and the others were formed mostly more recently than the uplifts of the Northern Rocky Mountains in Montana at the end of the Cretaceous Period. Hot molten rock at depth pushed upward the overlying sedimentary rocks beneath the western Great Plains, thus producing dome-shaped uplifts with younger overlying strata tilted away from the centers. Faulting producing offset rock layers has affected the rocks in these mountains to greater or lesser degrees.

The Judith Mountains have several exposed cores of Paleo-gene and/or Cretaceous granite. These core rocks, because of their greater resistance to erosion, stand higher today than the sedimentary rocks that they intruded and have about 2,000 feet of relief. The adjacent plains contrasts with the uplift in its low relief and grass cover.

NORTH
DAKOTA

Theodore Roosevelt National Park

G, P, A, E

The park in western North Dakota consists of two principal units, north and south (fig. 17A, sites 11a and 11b), the entrances to which are separated from one another by about 70 miles. The major geologic features in both units are badlands topography of exposed and deeply eroded bedrock, along with landslides and stream erosion landforms. The uplands surfaces of the North Unit were glaciated during the Pleistocene. Strata exposed in both units are parts of the Paleocene Fort Union Group. They consist mainly of coastal plain sandstones, siltstones, claystones,

Fig. 28. Coastal plain deposits and bison in the North Unit of Theodore Roosevelt National Park, North Dakota. Photo by author.

lignite coal, and beds of red clinker (scoria) left behind when some coal beds combusted. Some coal beds in the group outside of the park area in North Dakota and Montana are thick enough to be commercially exploitable. The coal is used to fire steam-powered electrical generating plants. Large, often nearly spherically shaped concretions, informally called "cannonballs," occur in some of the finer-grained mudstones and claystones. These concretions occur at rock exposures along park roads, particularly in the North Unit.

Fluvial (river) erosion over the last 2.588 million years (during the Late Pleistocene and Holocene) along the Little Missouri River and its tributaries in these parts of North Dakota has produced the badlands topography in the park units and adjacent lands.

Fossils in the formations include plant pollen grains, petrified wood, leaves, invertebrates, and vertebrates such as skeletal elements from turtles, fishes, and mammals.

Fort Union Trading Post National Historic Site

G, A

The buildings of Fort Union were erected on the first river terrace on the north bank of the Missouri River. The terrace is a relatively flat surface of the former floodplain of the river. There are good views of the valley from the edge of the terrace.

The river valley follows the edge of the maximum former extent of the last Pleistocene glaciation (Late Wisconsinan) across northern Montana (figs. 5, 15, 16). The river, fueled by huge amounts of water from the melting ice sheet, eroded the

Fig. 29. A flat, nearly horizontal river terrace (old floodplain) above the Missouri River at Fort Union, North Dakota. Many terraces are underlain by bedrock capped by river deposits. Photo by author.

valley. From time to time, the river deepened its valley, leaving behind remnants of old floodplains or terraces like the one here.

Hills to the north are capped by deposits of glacial till and erratic boulders left after the retreat of the last ice sheet. To the south of the valley, the hills are dotted with erratics deposited during melting of older Early Wisconsinan ice sheets that had advanced farther to the south.

SOUTH DAKOTA

Slim Buttes

G, P

I first saw these buttes on a field trip to the South Dakota plains about 110 miles north of Spearfish, at the northern end of the Black Hills dome, and have returned from time to time to muse about their origin. The buttes are mostly on federal lands in the Custer National Forest. The easiest access is along South Dakota Highway 20 either east from Buffalo, South Dakota, or west from Reva, South Dakota.

Fig. 30. Part of Slim Buttes, South Dakota. Note the color changes and changes in dips of bedding in the formations. This site illustrates the principles of superposition, original horizontality, and cross-cutting relationships. Photo by author.

From all directions, the site is a high area with the upper slopes covered in places by forests and parkland and in other places by badlands rock exposures. From a distance, the rock exposures look normal, with layered, nearly horizontal formations, but on approach some of the formations are tilted at odd angles to those above and below.

From oldest to youngest, the formations from the plains to the tops of the buttes are Upper Cretaceous Hell Creek and Paleocene Fort Union (our old Western Interior Seaway friends from farther north and west), the Upper Eocene Chadron Formation, the Oligocene Brule Formation, and lower parts of the Lower Miocene Arikaree Group. The latter three units are largely composed of volcanic ash fragments carried by winds onto the plains from eruptions to the west in the Great Basin.

The Arikaree beds are nearly horizontal. So are those of the Hell Creek Formation, but units in between are tilted and faulted. There are angular unconformities between faulted blocks and the nearly horizontal beds above. These structural features formed when the blocks moved along one or more detachment faults—a type of landslide within the Fort Union beds.

Visit the Badlands National Park (site 14), the Toadstool Park badlands (site 23), or the Chadron through Arikaree sequence in the southern Panhandle of Nebraska, southeastern Wyoming, and northeastern Colorado. It will be clear that, although the formations that you see here at Slim Buttes are present in those places, the unusual structural features are not. Muse on.

Badlands National Park

G, P, A, E

This South Dakota park is a great place to see vertebrate fossils in place, continental volcaniclastic sediments, ancient soils (paleosols), badlands topography, archaeological sites, and flora and fauna native to the Great Plains.

From nearly any vista in the park, it's easy to trace the stratigraphic succession because the rock layers have different colors and have weathered differently, depending on their hardness.

Fig. 31. Part of Badlands National Park, South Dakota. The yellowish-brown and red paleosols in the foreground and middle ground are weathered parts of the Pierre Shale deposited in the waters of the Cretaceous Western Interior Seaway. Photo by author.

Starting from oldest (bottom) to youngest (top), the hillsides are Upper Cretaceous Pierre Shale, a marine shale that is dark gray to black when unweathered, yellowish-brown when weathered. In many places the Pierre is overlain by a red-colored ancient soil. On top are light, pastel-shaded, predominantly volcaniclastic mudstone, siltstone, and sandstone deposits of the Chadron, Brule, and Arikaree units noted at Slim Buttes (site 13). Ancient buried soils or paleosols are very common in these formations. They can be picked out because they are usually somewhat more brightly colored (usually red and green shades) than the normally tan and gray colors of the other formations. Modern soils with these colors form in very warm and moist environments like those in the southeastern U.S. states. That was what the climate was like here when these formations were originally deposited. Take a careful look at those. You may see them at other sites, including Toadstool Geological Park (site 23).

Mount Rushmore National Memorial

G

The memorial, southwest of Rapid City, South Dakota, is in the central Precambrian core of the Black Hills dome. The forests, the crowds of tourists, and the sight of the sculptures of four U.S. presidents may distract your eyes and thoughts from the important thing here, the geology. Focus on the rocks in the memorial where the sculptures are carved. The faces are carved

Fig. 32. The Mount Rushmore sculpture, South Dakota. The faces were carved in Precambrian granites intruded into older metamorphic schists. Carol M. Highsmith Archive, Library of Congress, Prints and Photographs Division, LC-HS503–4146.

into coarsely crystalline granite. To the left of the faces and below, the rock is different in appearance, layered and darker. This rock is metamorphic schist. A feeder dike of granite cuts across the schist on the lower side (and thus must be younger than the schist) and continues as the mass of granite into which the faces were sculpted. The principle of cross-cutting relationships explains the sequence of events here (fig. 12).

The Black Hills is a dome-shaped mountain formed by the same folding event at the end of the Cretaceous Period that produced the Rocky Mountains (fig. 8). Paleogene and Neogene intrusive igneous rock masses, particularly in the northern half of the Black Hills, are evidence that some activity continued through the Cenozoic Era, as are the hot springs for which the city of Hot Springs is named.

Wind Cave National Park

G, E

In the Black Hills, caves have developed in the Pahasapa (Madison) Limestone Formation by dissolution of the limestone. Wind Cave, located north of Hot Springs, South Dakota, is outside the Precambrian central core of the Black Hills, in the Mississippian Madison Limestone. You have not read about the Madison in this book since its mention at Giant Springs

Fig. 33. Boxwork on the roof of Wind Cave in South Dakota. Note the uncommon shape of these calcium carbonate deposits, which are found in abundance here. Wikimedia Commons, National Park Service of the United States of America, photo by Dave Bunnell.

(site 6) in Montana, but the limestone beneath that place has generally been beneath you from there to here. This limestone was deposited as skeletal calcium carbonate debris left behind during the decay of the soft bodies of dead marine algae and invertebrate animals and the subsequent cementation of this mineral into limestone rock.

Both Wind Cave and Jewel Cave (site 17), and, for that matter, probably all caves in the Madison Limestone beneath parts of the Great Plains, formed in the same manner over a long span of geologic time. The limestone was deposited in seas covering the North American midcontinent during the Mississippian Period. A subsequent drop of relative sea level left the formation above sea level. Then weakly acidic groundwater, moving along fractures and bedding planes in the rock, dissolved parts of the calcium carbonate to form a cavern system. This exposure and dissolution may have begun geologically shortly after the formation had been deposited. After the initial cavern system was formed, sea level across the Midcontinent repeatedly rose and fell, leaving behind the sequences of sedimentary Upper Paleozoic and Mesozoic rocks that cover the Madison Limestone in the Great Plains. During these events, the Madison cavern system would have been below the water table and completely filled with water.

The folding that produced the Black Hills Dome at the end of the Mesozoic lifted the tilted and fractured limestone above the water table. The cavern system allowed water from rainfall and snowmelt to pass through it. When runoff water reached the cavern system, the slightly acidic water slowly dissolved more of the limestone and carried it away. Cave formations of calcite (calcium carbonate) formed later when the evaporation of water carrying those minerals in dissolved form allowed them to precipitate out of solution onto the cavern walls.

No one knows how many miles of caves there are in the Madison Limestone beneath the Black Hills, nor whether the caves are connected underground.

Wind Cave is best known for two features: the winds that blow through the narrow, natural cave entrance, which are caused by the difference in air pressure inside and outside the cave; and the occurrence of "boxwork" cave deposits of calcium carbonate shaped like open boxes that are found in abundance here but are not common in other caves. The stalactites, stalagmites, columns, and other features that you may think of as typical cave deposits are rare here.

The full Black Hills suite of rock types and formations occur within the park, including Precambrian granite, pegmatite and schist, Paleozoic and Mesozoic sandstone, limestone, shale and gypsum, and other rock types. Be sure to explore as much of the park as you can.

Jewel Cave National Monument

G

Jewel Cave is located about 15 miles west of Custer, South Dakota. Like Wind Cave (site 16), it was formed in parts of the Mississippian Madison Limestone. Both have some of the longest known explored cave passageways in the world. Both formed in generally the same way, but Jewel Cave has significant amounts of gypsum (calcium sulfate), indicating that some additional geologic processes were active here and probably not at Wind Cave during cave development. These additional processes perhaps included greater effects produced by water

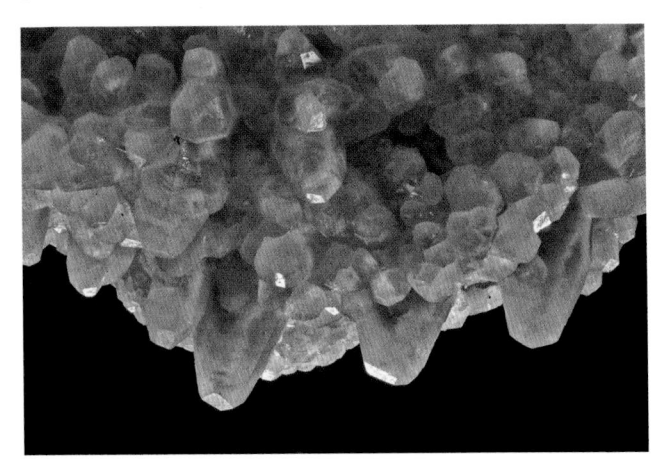

Fig. 34. "Jewels" of calcite spar crystals in Jewel Cave, South Dakota. Wikimedia Commons, photo by Dave Bunnell.

from thermal springs heated by magmas deep beneath the surface in that area.

Jewel Cave has some lesser amounts of boxwork as well as stalactites and other forms usually thought by the public to be typical of caves, but it is known for its huge areas of cave walls covered by blankets of calcite spar crystals and for the presence of gypsum crystals. I found these to be most wonderful to see and marvel over. Clearly something was geologically different than at Wind Cave for these kinds of cave deposits to predominate here and not there.

A word of warning: the tour is far more rigorous than the Wind Cave tour, with descents and ascents on stairways of hundreds of steps, even on the short tours. The trip is worth it if you enjoy walking through caverns and seeing beautiful cave formations, but do not go if you are not up to a long, hard walk through dark passageways.

Hot Springs Mammoth Site

G, P

Caverns can form in any place underlain by limestone. Where the limestone is at or near the land surface, sinkholes can form if the cavern roof is near the surface and other conditions are right. That's what happened beneath an area now located adjacent to a housing development in Hot Springs, South Dakota. In 1974 an equipment operator leveling land for a housing development found fossil mammoth bones. My longtime colleague Larry Agenbroad was called in and found bones from several

Fig. 35. Mammoth skeletons excavated in place, Hot Springs Mammoth Site, South Dakota. Mammoths slipped into and drowned in an ancient lake in a sinkhole in limestone formations. Photo by author.

more mammoths. The owner sided with science and the site was saved. Since the initial excavations, the digging and fossil preservation have continued. A large building now contains the excavated area, a gift shop, and a small museum with mounted fossils and other displays.

To date, parts of more than 60 mammoth skeletons have been excavated from the sinkhole deposits, and more may still be found. Most are Columbian mammoths, but a few are woolly mammoths. Remains of bear, camel, wolf, llama, coyote, small mammals, fish, amphibians, birds, freshwater snails, and clams have also been collected. More species may still be present, entombed in the thinly layered deposits that formed in the lake that partly filled the sinkhole at the time of the animals' deaths.

Petrified Forest, Black Hills

G, P

This site is located just east of Piedmont, South Dakota, about 15 miles north of Rapid City, on top of the Lower Cretaceous Dakota Sandstone hogback. Here the rocks of the formation tilt (dip) to the east. Pieces of petrified wood are frequently found on the Great Plains, especially in river gravel. It is rare,

Fig. 36. Part of an upright petrified tree stump in a modern pine forest in the Black Hills of South Dakota. The fossil stump is rooted in Lower Cretaceous sandstone beds. Photo by author.

however, to see whole fossil logs in place because the wood usually rotted away before the cell walls were filled with precipitated minerals. This place provides such a rare sight.

Enjoy walking around the site and seeing the fossils preserved in place along trails running through pine forest. The forest sits on the Dakota Sandstone Formation, another coastal plain deposit laid down along the western margin of the Western Interior Seaway during the Cretaceous Period. Visualize the coastal plain in Louisiana or southeastern Texas, with its rivers and forests, as you walk the trails. That area is similar to the ancient environment of this place.

Big Bend of the Missouri River

G

I lived for decades in Nebraska and never noticed on maps or read about this extraordinary entrenched meander of the Missouri River south of Pierre, South Dakota. It is perhaps the largest and finest example of such a meander that I have seen in the Great Plains. Today it is partly inundated by the Lake Sharpe Reservoir impounded behind the Big Bend Dam. The

Fig. 37. Part of a Missouri River–incised meander in South Dakota. The river curves on the upper left and right of the image and continues behind the viewer. The bedrock is Cretaceous Pierre Shale, a marine deposit, with some river gravels capping the surface in places. Photo by author.

site can be seen from the eastern side of the Missouri River on the Crow Creek Indian Reservation, but it is best viewed from the western side on the Lower Brule Indian Reservation. On the western side of the river valley south of the narrowest part of the meander neck, a trail leads to the top of the divide ridge at the Narrows Area Overlook. Follow all tribal laws and rules when visiting.

The curve of the Big Bend meander is about 30 miles long. The narrowest point of the meander neck is only about 2,200 feet across. You can stand there and visualize the boats of the Lewis and Clark expedition in 1804 traveling all those miles around the curve to advance only 2,200 feet farther north along the Missouri.

Entrenched meanders form when rivers flowing on fairly level ground begin to cut into the sediments and rocks beneath their valley floors due to uplift, sea level lowering, or combinations of the two. In this case, after the last ice sheets melted from the area, the Earth's crust, which had been depressed by the weight of the ice, bobbed back up (rebounded). This uplift may have been responsible, at least in part, for the river's down cutting.

Devils Tower National Monument

G, E

Close Encounters of the Third Kind, here we come!

The monument is about 27 miles northwest of Sundance, Wyoming. Devils Tower is located on the northwestern flank of the Black Hills. As you approach, the land is fairly flat and mostly forested or being used as cattle pasture. The rocks here

Fig. 38. Devils Tower in Wyoming. Note the polygon-shaped columnar joints in igneous rock. Because it is harder than sedimentary rock, igneous rock has intruded to form this erosional remnant, which towers high above adjacent lands. Photo by author.

are Upper Cretaceous sandstone, siltstone, shale, gypsum, and limestone deposited in or along the shores of the Western Interior Seaway. The tower is composed of intruded gray, coarsely crystalline igneous rock that is harder and, therefore, more resistant to erosion than the adjacent, slightly dipping sedimentary rocks. The rock is polygonally jointed, the nearly vertical surfaces of the joints producing the groove-like appearance of the tower.

The Devils Tower intrusion formed at or somewhat after the end of the Mesozoic Era's Cretaceous Period and the end of the dinosaurs. Geologists differ as to whether it is a remnant of a volcano that reached the surface of the Earth or a small intrusion of magma that never made it to the surface. This intrusive event may have occurred when the other intrusions on the north side of the Black Hills occurred. In any event, the rock is impressive with its wonderfully large, column-like jointing.

Pine Bluffs Archaeological Site

G, A

The first rest area in Wyoming on I-80, just west of the Nebraska border, is a nice place to view a scenic High Plains site with pine forests growing on the bluff side. Lodgepole Creek flowing out of Wyoming has carved a valley into the Ash Hollow Formation and underlying Whitney and Orella members of the Brule Formation here. The Ash Hollow Formation, the youngest

Fig. 39. Exposed historic materials in talus and alluvium deposits, Pine Bluffs Archaeological Site. Official Website of Town of Pine Bluffs, Wyoming, http://www.pinebluffswy.gov/index.asp.

formation of the Ogallala Group of Miocene age, was deposited mostly by rivers flowing from the northern Front Range and the Laramie Range of the Rocky Mountains. The Brule Formation of Oligocene age is mostly volcaniclastic debris. We have seen this formation at several other sites, including Slim Buttes (site 13).

All of the formations are gently tilted to the east at Pine Bluffs. Many of the upper Ash Hollow beds are well cemented and resistant to erosion, thus producing a westward-facing escarpment that has been used by humans for shelter and as a water source on and off for more than 10,000 years.

Windows of the Past archaeology site is a short walk on a paved trail from the rest area parking lot. The University of Wyoming runs the site, which is covered by a metal building. Artifacts on exhibit range from Paleoindian stone tools up to trash from the late 1800s to the early 1900s. Personnel from the university's Anthropology Department also run the High Plains Archaeological Museum in the town of Pine Bluffs. The site building is not always open, so check ahead if you plan to visit. Even if the building is not open, you can walk around the park area and see the trees, wildlife, stream drainages, and shelter that have attracted people here for more than 10 millennia.

NEBRASKA

Toadstool Geological Park

Over the years since I first visited this park, part of the Oglala National Grasslands, managed by the USDA Forest Service, the gravel access road has been much improved, but parts can still be difficult to drive on when the road is very wet. The park entrance is about 15.6 miles northwest of Crawford, Nebraska. The parking area and campgrounds are about 1.4 miles west of the entrance, and the park is open all year. Trails in the park

Fig. 40. Badlands in Toadstool Geological Park, Nebraska. The red and yellow colors are developed on weathered marine Pierre Shale. Parts of younger volcaniclastic beds are exposed in the middle ground and background. Photo by author.

are unpaved and also difficult to traverse when wet. The park's name comes from the toadstool-shaped masses of rock that occur there. These usually have a cap layer of sandstone resting on top of claystone or mudstone that erodes more readily, thus forming the characteristic shape until the capping rock topples. Many of the best of these, including one the size of a Volkswagen Beetle, have already toppled.

You can see, walk over, and touch parts of some of the same Paleogene formations that are exposed at Badlands National Park in South Dakota (site 14), but I think that this site is more intimate because you can get closer to the rocks. Here there are also fossil trackways of birds, rhinos, camels, and other vertebrates as well as invertebrates exposed on the upper surfaces of sandstone beds in several places. The Pierre Shale and overlying Chadron and Brule Formations are all faulted in places. Although somewhat distant from the Black Hills, the park area was still under the influence of the tectonic forces that produced the Black Hills Dome and later structural features.

Hudson-Meng Bison Kill Research Center

G, A, E

In 1954 two local ranchers discovered this site, located about 18 miles northeast of Crawford on the Oglala National Grasslands. My longtime colleague Dr. Larry Agenbroad (who worked at site 18, the Hot Springs Mammoth Site) and his students excavated here from 1971 to 1977. Work continues today, protected by a metal all-weather building.

Larry and his students found skeletal elements from at least 120 bison and associated Paleoindian stone points. Radiocarbon dates in the range of about 10,000 to 9,500 years before the present have been determined from bone and charcoal in the alluvium in which the materials were buried.

Fig. 41. Bison fossils buried in alluvium at the Hudson-Meng Bison Kill Research Center near Crawford, Nebraska. Photo by author.

Pine Ridge

G, P, A

Toadstool Geological Park (site 23), the Hudson-Meng Bison Kill (site 24), and Fort Robinson State Historical Park are all on the flanks of the Pine Ridge Escarpment, the break between the Unglaciated Missouri Plateau and the High Plains sections of the Great Plains Province (fig. 4, sections 2b and 4; appendix 1). You can find excellent views of this important physical and structural break on the way to these places, but

Fig. 42. View of the Pine Ridge Escarpment (in background), looking west, south of Chadron, Nebraska. The ridge has been tilted and uplifted along faults related to the deformation of the Black Hills. The land in the foreground is in an unglaciated part of the Missouri Plateau section of the Great Plains. Photo by author.

another, and perhaps better, view is on the plains south of Chadron, Nebraska, as you approach Chadron State Park on U.S. Highway 385.

The rock formations that crop out along the escarpment from base to top are parts of the Paleogene White River Group (Chadron and Brule Formations) and parts of the overlying Arikaree (Gering, Monroe Creek, Harrison, and Upper Harrison Formations) and Ogallala Groups. The White River through Arikaree units are mostly composed of volcanic ash debris carried here by winds from ancient volcanic eruptions far to the west of Nebraska. There are also some nonvolcanic stream and river deposits in these formations. The Ogallala Group is mostly composed of sediments eroded from the Rocky Mountains in southeastern Wyoming and northern Colorado and deposited by rivers on the plains during the Miocene.

Faulting has uplifted the High Plains surface here with respect to the lower lands to the north of the escarpment. The faulting has affected the rocks up through at least the Arikaree Group, which must have occurred more recently than when those rocks were deposited because the faults have cut and offset the rock formations.

Agate Fossil Beds National Monument

G, P

If you enjoy looking at beautiful High Plains countryside, fine geology, and vertebrate fossils in place, this remote monument is a delight. The visitor center is located in far western Nebraska, east on a spur off of Nebraska Highway 29, a 56-mile-long highway from Mitchell to Harrison with no services (traveler beware!). Things have improved since I first visited the site decades ago. Then, the "highway" was gravel; now it is paved. Then the headquarters was a trailer building; now there

Fig. 43. Disarticulated Miocene rhino bones in place at Agate Fossil Beds National Monument in Nebraska. The bones are in river deposits of Paleogene age. Photo by author.

are houses for personnel and an excellent, if small, museum. Then the trails to the fossil sites were unimproved; now they are paved.

A walk of about a mile across the Niobrara River and uphill will take you to the excavations. There you can see some of the fossils in place and experience a High Plains environment that has not changed much since these animals lived.

The first vertebrate fossil finds were made at Agate in 1885 by James and Kate Cook, who two years later bought the ranch on which the finds are located. Over many years, the Agate Springs Ranch hosted teams of vertebrate paleontologists from the University of Nebraska, the Carnegie Museum, Yale University, Harvard University, the Philadelphia Museum, the University of Michigan, the Field Museum, the American Museum of Natural History, and other institutions. The area continues to attract researchers today.

Agate has fossils of Early Miocene mammals that are not found in many other locations on the Great Plains or elsewhere. Fossils collected here are exhibited in museums in Washington DC, London, and other major natural history museums around the globe.

The vertebrate fossils excavated from the monument include ancient rabbits, beavers, camels, horses, rhinos, and deer, as well as strange beasts that have no living relatives, such as several species of oreodonts (sheep-like creatures), entelodonts (animals that looked like giant hogs), chalicotheres (resembling big horses with claws), and carnivores, including the extinct form called a beardog. Sediment-filled, corkscrew-shaped burrows of an ancient beaver are locally common and often well preserved.

From oldest to youngest, the formations exposed in the park are the Monroe Creek, the Harrison, and the Upper Harrison (or Marsland Formation) beds of the Arikaree Group. All of these units are composed mainly of volcaniclastic debris, some major parts of which were re-eroded by Early Miocene rivers elsewhere and deposited by these rivers in the monument area. The vertebrate fossils noted above come mainly from the Harrison and Upper Harrison units.

Scotts Bluff National Monument

Located in the Nebraska panhandle, Scotts Bluff is an erosional remnant once part of the Wildcat Hills, which are visible looking south. The larger Wildcat Hills are in turn erosional remnants of parts of the High Plains that are today being washed away by the North Platte River, just to the north of

Fig. 44. Scotts Bluff, east side, near the city of Scottsbluff, Nebraska. The top of the White River Group and base of the Arikaree Group are at the area marked by a color change and break in the slope about halfway up the butte. The rocks are mostly composed of cemented and uncemented volcanic debris carried by winds from eruptions to the west of the Great Plains. Photo by author.

the monument; Pumpkin Creek, to the south of the Wildcat Hills; and their tributaries.

This butte, rising more than 800 feet above the floor of the North Platte Valley, served as a signpost for Native peoples and people headed west more recently. The area has a rich history that is chronicled in displays in the visitor center museum. Here you can stand and walk in ruts made by wagons traveling the Oregon Trail. There are hiking trails as well as a paved road to the top, where you can see breathtaking views of the North Platte River, its valley, and the Wildcat Hills.

The rock strata exposed in the bluff face are parts of the Oligocene Brule Formation and the Miocene Arikaree Group (Gering and Monroe Creek–Harrison Formations). All of these units, just as farther to the north at Badlands National Park (site 14), Toadstool Geological Park (site 23), Pine Ridge (site 25), and Agate Fossil Beds (site 26), are composed mostly of wind-carried debris from eruptions of volcanoes in the Rocky Mountains and mountains farther west. The Brule is generally finer grained and holds vertical or near-vertical slopes well, but landslides do occur in both the Arikaree and the Brule. The Monroe Creek–Harrison exposed in the upper parts of the monument and along the trail down the butte is mostly weakly cemented, or uncemented, volcaniclastic sands with local strong calcium carbonate cement that forms elongated, gray, pipe-shaped, hard rocks called pipy concretions. These concretions are typical of the formation and are usually horizontal in orientation.

Drive south on Nebraska Highway 71 to the Wildcat Hills Recreation Area and Nature Center for outstanding views of the hills and of both the North Platte and Pumpkin Creek Valleys. The recreation area also has some exposures of the Miocene Ogallala Group on the higher spots for you to see. There are hiking trails and picnic areas as well. Standing at the top of the Wildcat Hills or on top of Scotts Bluff, imagine the huge amount of water and wind erosion that has gone on since the start of the Pleistocene 2.588 million years ago, when the North Platte River and Pumpkin Creek began to flow across this region and to erode out the their present valleys. The amount of sediment eroded away and carried to the Gulf of Mexico from this area has been vast.

SITE 28

Chimney Rock National Historic Site

G. P

This site was another major landmark along the Oregon Trail. An erosional remnant of the Wildcat Hills just to the south, it is wearing away rapidly due to the combined actions of wind, water, freezing, and thawing year after year. The site is owned by the Nebraska State Historical Society and includes a nice visitor center with a small museum.

Fig. 45. Part of the Wildcat Hills of Nebraska; Chimney Rock on right. Lower Whitney volcanic ash is the more cream-colored layer exposed toward the base of the rock and the bluffs. Ash underlies much of the Nebraska Panhandle and southeastern Wyoming. Its volcanic source is the ancient volcanic complex along the Utah-Nevada border. Photo by author.

Chimney Rock is about 325 feet high from base to top. The rock base is in our old friend the Brule Formation. The overlying parts of the mostly volcaniclastic Arikaree Group begin where the slope abruptly increases and the colors of the rocks change from shades of tan to shades of gray. The parts of the group preserved here are the basal Gering Formation and a small, and increasingly smaller, bit of the Monroe Creek Formation.

Volcanic ash deposits in both the Brule and the Gering have yielded dates of about 31.8 and about 30.6 million years ago for the Whitney Member of the Brule and about 28.3 million years ago for an ash in the Gering. The prominent white deposit in the Brule near the base of the rock, called the Lower Whitney Ash, averages nine feet thick across the Nebraska Panhandle and is perhaps the most widespread single continuous ash bed in the region. The source of the ash was a volcanic complex located along what is now the Utah-Nevada border.

Courthouse and Jail Rocks

G, P

Owned by the Nebraska State Historical Society, these buttes are the easternmost obvious remnants of the former extent of the Wildcat Hills. From base to top, the formations exposed in the westernmost and highest of the two rocks, Courthouse Rock, are the same as those exposed in Chimney Rock, namely the Whitney Member of the Brule overlain by the Gering Formation and parts of the Monroe Creek–Harrison Formations

Fig. 46. Looking north toward the erosional remnant buttes called Courthouse Rock (*left*) and Jail Rock (*right*) from Nebraska Highway 88. Photo by author.

of the Arikaree Group. The top of Jail Rock is lower and there is no Monroe Creek–Harrison there.

Both rocks can be approached on a gravel road. Carefully follow the beaten trail from the parking area to the base of the escarpment and the start of the Gering Formation above the Brule. Walk up the slope to see the several sets of sandy stream deposits, one on top of another, each grading upward into finer-grained eolian silt deposits in the Gering. While you are at this point be sure to look to the south across the valley of Pumpkin Creek to the distant escarpment edge of the Cheyenne Tableland. That tableland is underlain by parts of the Ash Hollow Formation overlying the Brule Formation, like the rock sequence at Ash Hollow State Historic Park (site 30), with few or no parts of the Arikaree Group rocks present between the two. Looking north across the wide North Platte River valley, you can see the escarpment edge of the southern side of the Box Butte Tableland, underlain there mostly by thick Monroe Creek–Harrison beds, some Gering Formation, and Whitney Member of the Brule. The Ash Hollow Formation in that area is usually thin or absent. The Monroe Creek–Harrison beds exposed in the upper parts of Courthouse Rock have pipy concretions typical of the formation.

The High Plains here was built up first by deposits of volcanic debris and later by mostly river-transported sediments deposited in ancient river valleys that extended out onto the plains east of the Rocky Mountains. Rivers shifted their positions throughout the Cenozoic, producing a series of valleys with fills of different age ranges. This site and the others nearby are great places to see these different kinds of deposits.

SITE 30

Ash Hollow State Historical Park

G, P, A

Try to visit Windlass Hill at the southern end of this Nebraska park when there are few other people around. After climbing to the top of the hill and standing in the ruts made by the wagons of people traveling west on the Oregon Trail, you, like me, may be able to conjure up a vision of those folks and their wagons and teams working their way down the hill to the floor of the

Fig. 47. Aerial view of Windlass Hill at Ash Hollow State Historical Park, Nebraska, with wagon ruts (*center left*) and discontinuous Ash Hollow Formation (Lower Miocene) exposures on slopes. The Ash Hollow Formation is mostly sand and gravel, some cemented, eroded by ancient rivers from the Rocky Mountains to the west and southwest. Photo by author.

valley of Ash Hollow Creek. This comes to me more easily on a foggy day. I have taken groups of my geology students and other interested people to this spot and the rest of the park many times since the early 1970s.

You might wonder, as I did, why the leaders of the wagon trains picked this spot to cross from the South Platte Valley to the North Platte Valley. It certainly is not the easiest place for a crossing, what with the steep Windlass Hill slopes to navigate. The reason seems to have been water availability. There are perennial springs near the mouth of Ash Hollow, clean and potable water sources that do not occur anywhere else along the south side of the North Platte Valley for many miles, except for one nearby, hard-to-reach site to the east. Water has drawn people to Ash Hollow on and off for at least the last 9,000 years, as evidenced by the archaeological record preserved in Ash Hollow Cave, a rock shelter near the northern end of the park.

I made geologic maps of the park and adjacent lands in the mid-1970s and co-wrote a field guide to the park. The oldest rocks in the park are part of the Oligocene-age Brule Formation that occurs at several of the sites described above. The exposures of this siltstone formation, composed mostly of volcanic ash, are all found at the northern end of the park, generally in steep bluffs and escarpments along the Ash Hollow valley.

There is a major unconformity at the top of the Brule that slopes rapidly to the south. The Ash Hollow Formation of the Ogallala Group of rocks occurs above the unconformity. This Late Miocene formation is dominantly composed of river deposits with some lake deposits and some volcanic ash lenses. The vertebrate fossils found in these beds are of types like those found at Ashfall Fossil Beds (site 34; see that site for types). Fossil insect and animal burrows and root traces and "seeds" of grasses, forbs (broadleaf herbs), and hackberry trees occur in the formation as well. Remnants of sand and gravel, left behind by the South Platte River when it crossed this area in the Pliocene about 2.6 million years ago, make up the Broadwater Formation, which caps the Ash Hollow Formation south of the park. These river deposits, in turn, are covered by thick Pleistocene and Early Holocene wind-deposited silts called loess.

The park and adjacent areas give you history, archaeology, paleontology, and geology all rolled into one small area. A visit is well worth the price of admission.

Sand Hills Region

G

The Sand Hills region is characterized by wind-transported sand piled up into large dunes of seven types. This region is the largest sand sea in the western hemisphere, covering some 20,000 square miles. Some of the dunes are tens of miles long and have heights of 300 feet or more. The sand deposits were derived from re-erosion of underlying Miocene through Quaternary river deposits upwind, that is, northwest of the dunes. Dune formation occurred more than once during the last 10,000

Fig. 48. The Nebraska Sand Hills (*background*) and a groundwater-fed lake. These hills are vegetated sand dunes formed principally from about 9,700 to about 7,000 years ago. Photo by author.

years (Holocene), but the longest period of activity appears to have been from about 9,700 to about 7,000 years ago.

Today the dunes are stable, covered by grasses that provide pasture for cattle. Lakes abound between the dunes in certain areas where the water table is high, attracting large numbers of migrating and indigenous species of birds. The greater part of the High Plains aquifer lies beneath the Sand Hills Region in Nebraska. One of the major aquifer systems in the United States, this is a huge water resource for Nebraska and states downstream.

I did not pick a particular site for you to visit, but below suggest several places to see how abrupt the change to dunes can be as well as the nature and variety of dune forms. One of the easiest places to see the abrupt change from the fairly flat High Plains to the Sand Hills is along Nebraska Highway 2 east of Alliance, Nebraska (fig. 10). Continuing along Highway 2 you drive mostly along the low areas between very large and long dune crests. Take Nebraska Highway 61 or U.S. 83 generally north-south across the long axes of the dunes to get a feel for the changes in dune shapes and heights.

Snake River Falls

G

This wonderful perennial waterfall, located on the Snake River about 23 miles southwest of Valentine, Nebraska, is not nearly as high as Smith Falls (the highest falls in Nebraska) but compensates for that by having a huge discharge. In contrast to Fort Falls (site 33), Smith Falls, and other falls on the Niobrara River, the water here plummets over the well-cemented Caprock Member sandstone at the base of the Ash Hollow Formation, both deposited by ancient rivers.

Fig. 49. Snake River Falls, Nebraska, with a cut-bank exposure of part of the Ash Hollow Formation on the right. Photo by author.

Fort Falls and Fort Niobrara
National Wildlife Refuge

G, E

Most people, including some Nebraskans, probably would be surprised to find perennial waterfalls and cascades in the state. Actually there are a number of them, mostly along the south side of the Niobrara River in north central Nebraska. Almost all of the water comes from groundwater springs that issue

Fig. 50. Fort Falls, Nebraska; really a cascade over a cliff cut into the volcaniclastic Rosebud Formation (Oligocene). Photo by author.

from the High Plains aquifer. The Niobrara River has eroded its valley below the level of the water table in this area, and groundwater flows out of the aquifer in the form of springs and seeps. The sediments and sedimentary rocks of the regional aquifer are thicker and far more extensive in areas south of the Niobrara Valley than to the north and thus contain more groundwater to feed the springs, seeps, and rivers of the area than do those to the north.

Fort Falls is close to U.S. Highway 20 and the city of Valentine, Nebraska, and the headquarters of Fort Niobrara National Wildlife Refuge. This cascade is located on the refuge and can be reached easily. A cascade is a kind of shortened rapids where water descends over steeply slanted rocks, rather than falling freely over a rock ledge as in a falls. The drop here is only about 40 feet, but the site is scenic and has the same general geologic controls as other, more remote, falls and cascades along the river.

From the river level upward in this area the exposed formations are part of the Oligocene-age Rosebud Formation. Parts of two formations of the Ogallala Group, the Valentine and Ash Hollow of Miocene age, overlie the Rosebud, with the base of the Valentine unconformable above the Rosebud siltstones. The Rosebud resembles the Whitney Member of the Brule but is a finer-grained siltstone equivalent in age to the coarser Gering and basal Monroe Creek–Harrison Formations to the west. Natural exposures of the siltstone form vertical or near vertical slopes, while the generally uncemented sands of the overlying Valentine Formation have much gentler slopes. Rosebud siltstones, like their coarser-grained equivalent to the west, have major volcaniclastic contributions to the rock materials.

The groundwater coming locally from above Fort Falls feeds a tributary to the Niobrara River that plummets over the top of the Rosebud escarpment as a cascade. The source of the water is groundwater in the porous and permeable sandy sediments of the Ogallala Group and overlying Quaternary gravels and eolian sands.

Two other nearby places are interesting, both geologically and scenically: Smith Falls and the reach of the Niobrara River from Valentine downriver that is designated as part of the Niobrara National Scenic River. Smith Falls has the highest drop of any of the falls or cascades in Nebraska, at between 60 and 70 feet. The geologic units are the same here as those at Fort Falls, but there is more exposure. Local outfitters offer canoe rides along the river.

The Ogallala and younger deposits in the area have yielded abundant fossils of vertebrates and plants (petrified wood, fossil "seeds," etc.). Some of the best of these are on display at the University of Nebraska State Museum in Lincoln.

Ashfall Fossil Beds State Historical Park

This site, located in a valley tributary to the South Branch of Verdigre Creek in Antelope County, Nebraska, is a National Natural Landmark. About 12 miles east lies the border of the Pleistocene ice sheet that marks the eastern boundary of this part of the Great Plains. At the park, fully articulated skeletons of Miocene mammals, birds, and turtles occur in place

Fig. 51. Exposed fully articulated skeletons of a fossil horse (*center*) and rhino skeletons in the Rhino Barn, Ashfall Fossil Beds State Historical Park, Nebraska. The fossils are preserved in volcanic ash deposited in a small lake during deposition of the Ash Hollow Formation. Photo by author.

in the volcanic ash where they died. These include five species of horses, three of camels, one of deer, one of rhino, three of dogs, three of birds, and two of turtles.

My friend and colleague Mike Voorhies, now retired from the University of Nebraska, discovered the first fossil rhino in 1971 while doing reconnaissance paleontology with his wife, Jane, at the site, then on private land.

According to Mike, he began to excavate the site in June 1977 and obtained funding from the National Geographic Society for further excavations that took place there into 1979. With a crew of eight students and other helpers, he collected a number of vertebrate skeletons that are now housed in the University of Nebraska State Museum, located on the main campus of the University of Nebraska–Lincoln. By 1991 the land had been purchased by the state, and a new park was subsequently developed by the University of Nebraska State Museum and the Nebraska Game and Parks Commission. From an initial visitor center and small Rhino Barn covering part of the ash bed area, more buildings, walkways, and interpretive displays and a large expansion of the Rhino Barn have been added over the years. Through all this time, Mike, Jane, and numerous helpers have continued to excavate the site and to conduct more reconnaissance fossil surveying across the area.

The fossils are entombed in a lens-shaped volcanic ash deposit up to about 30 feet thick just above the basal Caprock Member of the Ash Hollow Formation. The ash was deposited in a pond over a period of days or weeks, initially as air-fall ash, followed later by erosion and transport of ash by water and wind off land adjacent to the pond. Other vertebrate fossils have been found at the site, both below the ash and above it.

The most recent geologic date on the volcanic ash is about 12 million years ago. This ash, which has also been found at several other sites in western Nebraska, came from an eruption of a Yellowstone Hotspot volcano on the Snake River Plain in southwestern Idaho.

How serendipitous were the geologic events that led to this wonderful site being preserved for us to see? Just consider. A supervolcano, located on today's Snake River Plain in Idaho, shot volcanic ash high into the atmosphere. It then happened to be carried eastward to northern Nebraska by favorable winds. Herds of Miocene mammals and groups of other animals happened to be in the area of a water hole on the plains when the ash began to fall. The animals died and were quickly covered by more ash and later by other deposits of the Miocene Ash Hollow Formation. The mid-Pliocene ancestral Platte River (fig. 16) did not deepen its valley in this area and erode away the fossil-bearing beds. The western edge of the Pleistocene ice sheet (fig. 15) stopped a few miles short of moving this far west and destroying the fossils. Since the ice sheet melted, erosion of the tributary of the South Branch of Verdigre Creek has not proceeded to the point where all of the ash containing the fossils has been carried away.

SITE 35

Niobrara State Park

G, P

Niobrara State Park is located on the west side of the Niobrara River at its confluence with the Missouri River. This site and Ashfall Fossil Beds (site 34) are the easternmost locations discussed on our journey. Both are very near to where I have drawn the eastern boundary of the Great Plains Province (figs. 4, 17). The rocks beneath the park are Cretaceous shale

Fig. 52. Part of the new Niobrara State Park (*foreground*) in Nebraska, with the Niobrara River and the abandoned former park site just above center. The park grounds above the river are underlain by weathered Pierre Shale on top of Niobrara Formation chalk. Photo by author.

beds of the Pierre Formation, one of many Western Interior Seaway units deposited in that sea, all of which are subject to landslides. You can see many active and inactive landslides in the park and along the adjacent roadways and countryside. Below the Pierre Shale, close to the river level, there are natural and man-made exposures of the upper beds of the Upper Cretaceous Niobrara Chalk Formation, another of the Western Interior Seaway units. Landslides also occur in parts of this formation.

This area shows well the unintended consequences of human activities on the natural and built environments. In 1930 the town of Niobrara, Nebraska, was laid out on the south side of the Missouri River floodplain at the base of bluffs underlain by the Pierre and Niobrara Formations. A park was situated on an island of the Niobrara River just upstream from the river's confluence with the Missouri. That year the land was transferred to the State of Nebraska and became Niobrara Island State Park.

The town might flood if the Missouri did, but was unlikely to be involved in a landslide of the adjacent hill slopes. However, after major floods in 1942 and 1943, the federal government decided to build dams on the Missouri to control future floods. One of these, Gavins Point, was built by the U.S. Army Corps of Engineers between 1952 and 1957. The reservoir behind the dam filled up close to the town of Niobrara. That might not have seemed so bad at the time, but it also raised the water table beneath the floodplain. Consequently, basements flooded, and mostly dry ground became marshy. The sand-transporting Niobrara River began to deposit sediment in its channel, thus building up the channel and flooding the park. As a result, the town was relocated up onto the bluffs area, and the state park was moved to the west side of the Niobrara River on the bluffs. Both of these spots are now more susceptible to landslides than formerly and are no longer located above good aquifers. Furthermore, the reservoir behind the dam is shallow and rapidly filling with sediment. This is an outstanding area to see the visible impacts of human-induced changes in land use.

Harlan County Lake

G, P, A

I spent considerable time studying the rock strata, structures, landslides, and fossils at this reservoir in south-central Nebraska from 2000 through 2002. Most of the best bedrock exposures are on the south side of the reservoir because the Republican River began to initially erode out its valley to the north of its

Fig. 53. A normal fault cutting across dark-colored Cretaceous Pierre Shale Formation (*left*) and light-colored Niobrara Formation (*right*) at Harlan County Lake, Nebraska. The uplifted side is on the right. This is an example of cross-cutting relationships. The faulted rocks lie below an unconformity overlain by unconsolidated Holocene-age deposits. Photo by author.

present position and gradually shifted to the south through time, eroding away bedrock and leaving river deposits behind on the north.

The rock exposures visible along the cliffs on the south side of the reservoir are the only sites that I know of in the Great Plains where it is possible to view so many earthquake faults. If the water level is low, as it was during the drought years of 2000 to 2002, you can walk almost all of the reservoir shore and see the faults. If the pool level is high, you can still see some of them from the lake access roads along the south shore.

Parts of the Upper Cretaceous Niobrara Chalk and Pierre Shale Formations and small remnants of the Ogallala Group are exposed in the bluffs along the south shore. Faults offset both the Niobrara and the Pierre, but not the Ogallala. The bluffs and their rock faces are kept fresh by erosion from the undermining actions of breaking waves on the bluff faces during times of high pool levels.

There are at least 17 faults along the shore. Most of them are high-angle normal faults, but there are some grabens and a few minor reverse faults, too. I estimate maximum vertical movements on some of these at more than 40 feet, as evidenced by the change in altitude of the top of the Niobrara from one side of the fault to the other.

I suspect that there are faulted strata like these in other parts of the Great Plains. Most areas, however, are usually covered by geologically younger surface deposits that mask their presence. In this case, the faulting is probably caused by renewed post-Cretaceous uplift along the crest of the Chadron-Cambridge arch (an upfold in pre-Cenozoic rocks beneath Nebraska from Chadron to Cambridge) that runs beneath the area. If you are interested in the kinds of rock structures that are not often visible, this is definitely a place to go.

KANSAS

SITE 37

Monument Rocks

G, P

My visit to this site and the following one, Castle Rock, Kansas (site 38), in September 1975 was serendipitous. At the time I was teaching geology at Doane College in Crete, Nebraska. One of my colleagues, Dan Deines, told me that a friend who farmed near Wakeeney, Kansas, had literally run across what he thought were big fossil elephant bones while tilling his wheat field. The friend wanted someone who knew about such things to take a look, evaluate the find, and remove the bones intact, if possible, to be preserved. Thinking that this would be a good experience for my undergraduate geology students but not knowing much about excavating such bones, I contacted

Fig. 54. Monument Rocks, Kansas. An erosional remnant of Cretaceous Niobrara Chalk. Wikimedia Commons by Brian W. Schaller.

my friend George Corner at the University of Nebraska State Museum. Dan, George, the students, and I loaded up collecting gear and headed to Wakeeney. During the day, we excavated and jacketed the bones of a mammoth, some of which were loaded in the van and taken to the college for preparation by the students. The remaining bones were left with the farmer. Later in the day, we had enough time and energy to explore two nearby sites where buttes of Upper Cretaceous Niobrara Chalk stood as isolated sentinels above the plains.

We first visited Monument Rocks, erosional remnants from the Upper Cretaceous Niobrara Chalk Formation exposed along tributaries on the north side of the valley of the Smoky Hill River. This site can be seen from county roads. On private land (which may not be open to the public), it is a National Natural Landmark. In these remnants you can see the thin layers of chalk, roughly equivalent in age to the chalk of the White Cliffs of Dover in England. The chalk is composed mostly of skeletal debris from various species of marine algae and microscopic protozoan (foraminifera) shells. Larger fossils of clams, ammonites, sharks, bony fishes, and other species of ocean-dwelling invertebrate and vertebrate animals have also been found in this formation. This site is yet another deposit from the Cretaceous Western Interior Seaway.

Castle Rock

G, P

This Kansas site, like Monument Rocks (site 37), is on private land and may not be open to the public. It is another remnant of the Niobrara Chalk Formation and an alternative site to visit to see the same sorts of buttes that are visible at Monument Rocks.

Fig. 55. Castle Rock, Kansas. An erosional remnant of Cretaceous Niobrara Chalk. Photo by author.

COLORADO

Castle Rock

Castle Rock, in the Colorado Piedmont Section, is hard to miss seeing as you drive into the city of the same name from either the north or the south on I-25. The rock, now a city park called Rock Park, is a flat-topped butte east of the highway. To view the rock closely, take exit 182, park at the northwest trailhead in the park, and hike the trail to the top.

Whenever I see a butte of this sort with harder rock at the top forming a vertical escarpment, I immediately hypothesize

Fig. 56. Castle Rock at Castle Rock, Colorado. An excellent example of "inverted topography," with river deposits, once in the lowest spot on the land surface, today capping the highest hilltop in the area. Photo by author.

that this might be a place where topography could be inverted, that is, where a topographically low place on the land was located in the past.

The area has been studied by geologists on and off since the late 1800s. A USGS folio with a geologic map and photographs of the Castle Rock area by G. B. Richardson was published in 1915. The areal distribution of the several rock formations and rock types depicted on that map have not significantly changed since then. The formation capping the Castle Rock butte is part of the locally widespread Castle Rock Conglomerate Formation, an ancient river deposit of Late Eocene age that tops many of the highest hills in the area. The slopes below the conglomerate are underlain by parts of the Dawson Arkose of Paleocene age.

An ancient erosional surface or unconformity lies between the uppermost arkose (a kind of sandstone) and the conglomerate. This surface has the form of a river valley into which the river gravels, later cemented to form the conglomerate, were deposited. Modern stream and river erosion in the area has left behind large remnant areas capped by the conglomerate. If this description causes you to think back to the Cypress Hills Formation conglomerates in Alberta and Saskatchewan (site 3), you have been paying attention. At both sites, large rivers flowed from the mountains onto the plains in the past in places where today there are only uplands.

Not all of the flat-topped hills on either side of I-25 between Monument, Colorado, and Parker, Colorado, are capped by the Castle Rock Conglomerate. An igneous rock formation, the Wall Mountain Tuff (also called Castle Rock Rhyolite by local geologist Peter Laux), a volcanic flow deposit of somewhat older Late Eocene age, caps many of these. This cemented volcanic ash (tuff), emanating from the Sawatch Range, covered parts of the Late Eocene landscape. In some places the Castle Rock Conglomerate rests on top of the tuff and the conglomerate contains big pieces of the tuff, and so the conglomerate must be younger. Richardson reported and mapped this relationship back in 1915, also.

There are some structural adjustments by faulting and by broad, rather gentle folding that have affected the area, but these are usually not obvious. They have been recognized and reported since publication of Richardson's folio.

To explore and study the formations in detail, including the tuff, I recommend that you visit Castlewood Canyon State Park southeast of the town of Castle Rock. There are several hiking trails where you can walk in a safe environment on public land.

Section Boundary, Raton/Colorado Piedmont

G

Although it's been extensively studied, there is little agreement as to where to draw the boundary between the Colorado Piedmont and the Raton Section to the south. Is it the Arkansas River or just to the south of the Arkansas River? Or is it at or just south of Pueblo, Colorado? I looked at both areas and was satisfied that there was a great enough change in landforms and geology for either to be the basis of a boundary.

Fig. 57. Boundary between the Colorado Piedmont and Raton Sections of the Great Plains along I-25 south of Pueblo, Colorado. The fault crosses at about the bend in the highway. Photo by author.

I propose that the boundary be placed crossing Colorado I-25 at the prominent hill crest (fig. 57) between exits 71 and 74. South of this hill the land flattens out and seems distinctly different. The safest place to see this break, which also may be structural (faults are mapped crossing the area here), is on the access road on the east side of I-25 at exit 74. Turn south on the access road immediately east of the I-25 exit and drive south a short distance to where the road turns east. Stop in a safe place and look southwest to where I-25 comes upslope on a steep rise and can no longer be seen. That is where I drew the boundary. On the highway, the road is very steep here and requires any driver in his or her right mind to slow down, both ascending and descending the rise.

Comanche National Grassland

G, P, E

I had the opportunity in late October 2013 to visit parts of these grasslands as a participant on a field trip led by Steve Miller of the Western Interior Paleontological Society and Bruce Schumacher of the USDA Forest Service from La Junta, Colorado. If you are interested in seeing the area, I suggest

Fig. 58. Comanche National Grassland overview from east side of Forest Service Road 25 in Colorado, looking southwest. The rocks in the road cut are Cretaceous age, with lighter-colored Niobrara Formation (*top*) and Carlile sandstone and shale (*below*). Sandstone beds at the top of the Carlile are grayish-brown layers forming ledges; dark gray Carlile shale beds lie below the sandstone. Older Cretaceous sedimentary rocks underlie the plains beyond. Photo by author.

that you contact the staff at the La Junta Office of the Forest Service about which parts of the grasslands are publically accessible, current programs and tours of the grasslands, and what collecting, if any, is permissible. Among the attractions are large numbers of individual dinosaur tracks and trackways; fossil leaf impressions; marine fossil bivalves, ammonites, and other invertebrates; and fish bones and teeth.

The grasslands are in part of the High Plains Section of the Great Plains Province in southeastern Colorado and are divided into two separate units, the Timpas and the Carrizo, separated from one another by privately owned lands. My field trip included parts of the Timpas Unit south of La Junta and Rocky Ford. The bedrock exposed here ranges from the Dakota Sandstone of Early Cretaceous age up through the lower part of the Niobrara Formation of Late Cretaceous age. These beds are partially exposed in the valley sides of the northeast-flowing Purgatoire River. The river has cut down through some gentle up-arching of the Cretaceous and underlying strata.

The three sites that we visited on the trip were on the north-dipping cuesta north of the river. The beds of the Smoky Hill Member of the Niobrara are at the high points of the grasslands here, the older Greenhorn Limestone occupies the intermediate slopes, and the Dakota crops out in tributaries close to the river. These formations were deposited during one complete cycle of the rise and fall of the Cretaceous Western Interior Seaway (Dakota Sandstone, Graneros Shale, Greenhorn Limestone, and Carlile Shale) and during part of the subsequent cycle (Niobrara Formation).

Drive downslope from the Niobrara site toward the river to the Dakota site in Minnie Canyon. There are outcrops of cross-bedded sandstones and ripple-marked sandstones in the Dakota with well-preserved leaf impressions and footprints of carnivorous dinosaurs there. The sands, later cemented into sandstone, were deposited by rivers in a near-shore coastal plain setting.

Next, visit Vogel Canyon (Davis Canyon on older maps) to look at parts of the Greenhorn Limestone. Alternating shale and limestone beds were deposited during wet and dry climate cycles. The shale beds are composed mostly of land-derived muds, the limestone beds mostly of organically produced marine sediments combined with muds washed off of the adjacent lands and into the seaway. There are ammonites (extinct relatives of the chambered nautilus) in the limestone beds.

Finally, stop at the overlook on Forest Service Road 25 (fig. 58). There you can see on top the cream-colored marine limestone and chalk of the lower part of the Niobrara Formation above the beach and near-shore sandstone beds at the top of the Carlile Shale. The grayish-brown sandstone beds in the Carlile form layers below the Niobrara; below the sandstone is dark-gray marine shale in the Carlile. This section is well exposed along the west road cut on Road 25 downslope from the top of the escarpment, but it is not a particularly safe spot to view the rocks because of vehicle traffic. The safest area for viewing the sequence is northeast of the Park Service road near the edge of the escarpment. Ammonites, bivalves, and fish fossils occur in these beds.

SITE 42

El Huerfano

G

The Raton Section shows evidence of relatively recent volcanism. Traveling on I-25 south of Pueblo, Colorado, your eyes will be caught by El Huerfano, a dark gray to black, inverted cone-shaped butte that sticks up high above the adjacent land. The top is irregular but generally looks as though it has been cut off. There is a darker layer of steeply tilted rock crossing the west side. The butte has been interpreted as a volcanic neck, the feeder tube for molten rock moving to the surface volcano. The

Fig. 59. El Huerfano, near I-25 in Colorado. The igneous intrusion is more resistant to erosion than adjacent sedimentary rock layers. Photo by author.

darker layer crossing the butte indicates that something more than just a one-time formation of a volcanic neck happened here (remember the principle of cross-cutting relationships).

There is a parking area with explanatory signage off the northbound lane just before I-25 crosses Huerfano Creek, just south of a newly installed wind farm. I recommend using the pull-off for viewing.

While this is the first obvious volcanic feature south of Pueblo, there is also one (west of I-25 at exit 67) farther north that was mapped by R. C. Hills in the late 1890s.

SITE 43

Spanish Peaks

G

Southwest of El Huerfano and south of La Veta, Colorado, two large, igneous, intrusive bodies called the Spanish Peaks jut sharply out of the plains. The peaks are parts of the Spanish Peaks Wilderness, an area of about 30 square miles managed by the U.S. Forest Service.

Fig. 60. An intrusive dike with the north slopes of West Spanish Peak in the background. The intrusions of rocks of Colorado's Spanish Peaks and associated dikes formed in the latter parts of the Late Cretaceous to Early Paleogene mountain building that produced the Rocky Mountains. Photo by author.

The Spanish Peaks are usually not included in the Southern Rocky Mountains (Sangre de Cristo); most geologists place them in the Great Plains because the peaks are east of the faulted and folded mountain front. Although they have forms that resemble volcanoes, the Late Miocene magmas that produced the peaks were entirely intrusive and never reached the surface. You might think of them as two "blisters" that pushed up through or intruded into the overlying sedimentary rock sequence here. The sedimentary rocks are less resistant to erosion, and so the formations above and adjacent to the "blisters" (or stocks, in "geospeak") have been eroded more deeply, leaving the stocks as higher peaks. Nearly vertical and vertical dikes radiate outward from the peaks, some running for miles across the landscape. They resemble walls of dark-gray rock visible from the upland as you approach La Veta. South of La Veta, Colorado Highway 12 cuts through one of these dikes. The same kinds of intrusive structures are present at the Upper Missouri Breaks (site 7), but those here are easier to see and to approach for viewing.

You can understand why earlier researchers differed on where to place the Spanish Peaks when you travel west on U.S. Highway 160 from Walsenburg toward Fort Garland, crossing the boundary between the Great Plains and the Rocky Mountains. Another feature like the Spanish Peaks is present on the north side of the highway northwest of La Veta, but closer to the mountain front. Some mountains northeast of Fort Garland also have a similar look.

NEW MEXICO

Raton Pass

Raton Pass is a low spot in the Raton Mesa, along I-25 in New Mexico. The mesa is another erosional remnant, and even though it displays great topographic relief, it is still a part of the Great Plains.

One reason that the mesa has not been eroded away is because it is partly capped by very hard and durable lava flows. These have protected the mesa from the great erosion of the plains to the north and south. A second reason for the preservation is that no major rivers have flowed across this area during Quaternary times, while they have in the areas to the north and south.

You might wonder, as I have, what possessed travelers headed to Santa Fe from southern Colorado to go up and over such a big topographic obstruction, but you only have to drive around the mesa to come up with an answer that satisfies. There simply are no better, easier, or shorter routes, even though weather in the pass can be difficult during many times of the year.

The bedrock exposed in road cuts and natural exposures ranges from Paleocene at the pass to Upper Cretaceous lower down the slope. All of the formations are terrestrial deposits and many are from river deposition. The Cretaceous sequence includes some prominent coal beds that, if you are a passenger in your vehicle, you may glimpse during the hair-raising ride up or down the south side of the mesa. A safer and more serene view of Raton Mesa can be found at Sugarite Canyon State Park (site 45).

SITE 45

Sugarite Canyon State Park

G

This park is a safer place to view and study the Raton Mesa than trying to pull off of I-25 crossing Raton Pass. Traveling upslope from this New Mexico park's visitor center, the rocks are Cretaceous marine Pierre Shale overlain by coastal plain deposits of the Trinidad Formation. The Upper Cretaceous to Upper Paleocene Raton Formation includes more coastal plain and marginal marine sedimentary rocks, including coal beds, sandstones, and mudstones. The Cretaceous-Paleogene

Fig. 61. Upper Cretaceous Seaway marine and coastal plain beds capped by hardened Neogene basalt flows, Sugarite Canyon State Park, New Mexico. Photo by author.

boundary impact debris marking the dinosaurs' end is present in this section just as in the sequence at Grasslands National Park, Saskatchewan (site 5).

Above the Cretaceous sequence, hardened, black-colored lava flows form a protective cap. You can easily recognize these flows by their color and weathering pattern. The rock is resistant to erosion and forms vertical faces. The rocks also are vertically jointed. These basalts, in contrast to some of the earlier noted volcanic rocks in the Raton Section, are Late Miocene to Late Pliocene in age, about 8–3 million years old. You can see the basalts and exposures of parts of the underlying formations as you drive or hike around the park.

SITE 46

Capulin Volcano National Monument

G

Capulin is a very young geologic feature, having reportedly formed between 62,000 and 56,000 years ago. It is so fresh looking that a visitor might wonder if it will erupt again anytime soon. From New Mexico Highway 325 a paved road leads to a parking area at the top of the 1,300-foot-high cone. From there you can walk around the crater rim and see volcanic features for miles in all directions on a clear day. Look for volcanic cones of several sizes (fig. 9) and the higher front edges of hardened lava flows on the plains below. The views are absolutely spectacular. If you are a photographer, I suggest that you stay from dawn to dusk so that you can take pictures of the landscape and life changes throughout the day.

Fig. 62. Mount Capulin, a volcanic cinder cone of Late Pleistocene age in New Mexico. Wikimedia Commons, U.S. Geological Survey.

Clayton Lake State Park

G, P

The drive from Clayton, New Mexico, to Clayton Lake State Park takes you through hardened volcanic flows and cones. I recommend this park for close views of volcanic rock and sedimentary rock with dinosaur footprints.

Quaternary basalt flows cap the uplands on the southern side of the park and lake. Parts of the Upper Cretaceous Greenhorn Limestone Formation, Graneros Shale, and Lower Cretaceous Dakota Group sandstone and shale beds are exposed below

Fig. 63. A fossil dinosaur footprint in Lower Cretaceous Dakota Sandstone at Clayton Lake State Park, New Mexico. Photo by author.

the basalts on the pathway leading along the south side of the reservoir eastward to where the dinosaur tracks are visible.

Footprints of several types of dinosaurs are exposed on a bed of the Dakota Sandstone of late Early Cretaceous age. Both herbivore and carnivore tracks are present, with those of herbivores far more common. Tracks of a crocodilian have been recognized as well. The bedding surface on which the tracks are preserved is also ripple marked and mud cracked in places. Mud cracks would indicate that the surface sediment was exposed to the air at times.

SITE 48

Section Boundary, Raton/Pecos Valley

G

Part of the boundary line between the Raton and the Pecos Sections is structural. One such spot is in New Mexico on I-25 just south of exit 341. Here a steeply dipping hogback of Cretaceous rocks separates generally flat-lying Permian sandstone formations to the west from Cretaceous sandstone and other formations extending eastward. The safest way to see this road cut is from a frontage road on the south side of I-25, southwest of the exit.

Fig. 64. Steeply dipping Cretaceous Dakota Sandstone. View to the southwest along I-25 at the boundary between the Raton and Pecos Valley Sections of the Great Plains in New Mexico. The section boundary is clearly structural here. Photo by author.

Carlsbad Caverns National Park

G

Carlsbad Caverns National Park, New Mexico, is in the Pecos Valley Section of the Great Plains Province. The site is located topographically high in what is generally called the Guadalupe Mountains. The land is rugged. Some call the high spot "mountains"; however, there are no folds, faults, doming, or volcanoes that produce mountains. The land is dry here; cactus and other xerophytes are very common. The land does not look like other grass-covered parts of the Great Plains. Soils

Fig. 65. Permian limestone back-reef beds near Carlsbad Caverns National Park, New Mexico. These beds are part of a huge organic reef complex bordering an ancient deeper ocean basin. Photo by author.

are thin, and the Permian limestone bedrock is either at or just below the land surface.

While driving across the rolling desert terrain over those Permian limestone beds, it is hard to imagine what this area looked like in the Permian Period. Like most geological formations and landforms, these were also produced under vastly different environmental conditions than those at the site today. This parched land was once submerged beneath an ocean. The limestone beds are made up mostly of organically produced calcium carbonate grains and skeletal elements from calcareous algae, sponges, and other marine forms. These limestone beds were originally deposited in a southern part of the Permian Midcontinent Seaway, somewhat similar to the Cretaceous Western Interior Seaway, but much older.

The limestone formations were parts of a huge reef complex in a part of the seaway that stood high above the adjacent ocean floor from about 268 to 255 million years ago. Just as with the Cretaceous deposition, the Permian reef deposits were laid down as sea levels fluctuated and the region subsided. At any one time, from the reef top to the base of the reef flanks, the difference in water depth might have been as much as 1,000 feet. What is truly amazing to me is that this reef complex is standing high and dry today but still has the reef geometry and shape that you can see as you cross the region.

The question of when some of the limestone was dissolved to form the caverns has been much debated. The structure of reefs is porous. If sea levels dropped from time to time during the Permian, the exposed rock could have been partly dissolved by soil acids. However, the current opinion of Park Service geologists is that, unlike that process in other areas and geologic times, the major limestone dissolution took place here much later. During the Miocene Epoch, weak sulfuric acid solutions, formed when hydrogen sulfide was liberated from petroleum trapped in the limestone, combined with groundwater and moved through pore spaces in the rocks. Since the Pliocene, uplift of the region, weathering, and erosion have produced cave roof collapses due to water percolating through the cavern system. Because the caverns are connected to the outside surface atmosphere, evaporation has occurred, and cave formations like stalactites, stalagmites, columns, curtains, and other wonderful shapes have developed. The sulfur from the sulfuric acid combined with calcium and oxygen ions plus water to form gypsum varieties also found as crystal masses in the caverns.

The caverns, the fossil reefs, and the desert environments and life forms are all beautiful. This place is worth your time to visit, as is Guadalupe Mountains National Park, Texas (Site 51), part of the same Permian reef complex.

Blackwater Draw Clovis Culture Site

G, A

One of the most important sites for studying the early people of North America, Clovis is a place of special interest because it is so old by America's standards. It continues to be the focus of debate between the "Clovis First" researchers and those who think that humans had arrived in the Americas before the Clovis Culture appeared around 13,500 years ago. The Clovis Culture Site, located north of Portales, New Mexico, is open

Fig. 66. Exposed bones and artifacts excavated in lake deposits at Blackwater Draw Clovis Culture Site, New Mexico. Photo by author.

to the public. The Blackwater Draw Museum, a few miles to the east, is an excellent place to learn more about these peoples. Both are owned and operated by Eastern New Mexico University in Portales.

The dig site, in Upper Pleistocene and Holocene lake deposits, is housed in a metal building some distance from the entrance, across an abandoned gravel pit excavated into the underlying Ogallala Group (supposedly gravel, although there were few obvious gravel-sized rocks there on our walk). Clovis, Folsom, and younger archaeological layers with artifacts and bones of extinct and living species are preserved in the building for view. In Clovis and later times, people lived around the spring-fed lake. Today Blackwater Draw is only an intermittent drainage to the southeast.

TEXAS

Guadalupe Mountains National Park

G, P

Driving northeast toward Carlsbad Caverns on U.S. Highways 82/180 in Texas, you will pass El Capitan, the magnificent escarpment underlain by the Permian Capitan Limestone that forms the scarp, part of the reef complex that includes rocks dissolved in part to form Carlsbad Caverns (site 49). The rocks on view here are a fossil reef core. Visualize the whole area

Fig. 67. El Capitan at Guadalupe Mountains National Park in Texas, just south of its border with New Mexico. Reef core limestone forms the escarpment. The steeply dipping fore-reef beds on the left are composed of eroded reef debris deposited on basin slopes below the reef. Photo by author.

covered by the sea up to just above the top of El Capitan and imagine the huge drop-off into deep water in front of the reefs.

You can walk trails from the visitor center at the park head-quarters upslope toward El Capitan for a real workout. Check out the displays in the visitor center as well.

Llano Estacado Escarpment

G

Traveling eastward from Lubbock, you can find roads that pass down the eastern-facing escarpment of the Texas High Plains, or Llano Estacado. The escarpment is impressive from the ground but even more impressive from the air. Below the soils and the usually thin cover of Quaternary sediments, the first bedrock formation exposed is the top Caprock caliche of the Miocene-age Ogallala Formation. This calcium-carbonate

Fig. 68. Aerial view of the Llano Estacado caprock escarpment east of Lubbock, Texas, with the surface of the High Plains to the west in the background. Photo by author.

Fig. 69. White Caprock caliche (ancient soil) at the Llano Estacado escarpment with older Ogallala rocks below. Caprock is the topmost member of the Ogallala Formation (Miocene). Photo by author.

cemented ancient soil unit is very thick and forms a vertical or near vertical face in most places.

The Bridwell and Couch members of the Ogallala lie beneath the Caprock. Of these, the Couch has been interpreted as a sandy eolian deposit. I cannot recall any similar, clearcut and thick eolian unit in the Ogallala of Nebraska. This should not be surprising because ancient environments differed across the Great Plains, as do modern Great Plains environments from Canada to Mexico.

Enchanted Rocks State Park

G

Enchanted Rocks is located in the central core of the Central Texas Uplift, a rather symmetrical feature, the center of which is located on an east-west line between Georgetown and Eldorado, Texas. The center of the uplift is a dome-like structure of granites and metamorphic rocks overlain by tilted Paleozoic sedimentary rocks. The doming occurred at the end of the Paleozoic Era and may have been related to the tectonic plate collisions that formed the Appalachian and Ouachita Mountains at that time in geologic history.

Fig. 70. Exfoliated Precambrian granite at the top of Enchanted Rocks State Park, Texas, the site of an ancient uplift. Photo by author.

During the time of the Cretaceous Western Interior Seaway, the dome was a high area on the floor of the seaway, which was covered by Cretaceous marine limestone beds. The central core of the dome was then exposed by river erosion after uplift of the area west of the Balcones Fault.

The excellent exposures in the park are of granite, for the most part. The granite weathers into smooth, exfoliated sheets, leaving topographically high, dome-shaped landforms that resemble those at Stone Mountain, Georgia, and at Mount Monadnock, New Hampshire.

If you are able, walk up the trail to the summit of Enchanted Rock and view the adjacent lower granite domes and exfoliating granite sheets. Observe the more deeply weathered areas on the granite surface where rainfall collects in the lows, allowing plants and animals a tenuous hold on life. To the south, you can see higher hills capped by Cretaceous limestone beds dipping gently to the south away from the Precambrian core.

This is a great place to visit, but take plenty of water if you plan on hiking to the top or around the rocks. There is no drinking water available along the trails. When you are thirsty, "enchantment" loses its meaning.

Gault Archaeological Site

G, A

At the Gault Site northwest of Georgetown, Texas, though remote today, humans have lived fairly continuously for at least the past 14,500 years. It has a combination of unusual geology, ecology, and archaeology. If you want to visit the site, contact either the Gault School of Archaeological Research in San

Fig. 71. An excavation pit at the Gault Archaeological Site in Texas. Note the stratification of Holocene stream deposits in the walls of the pit, indicated by color changes. Photo by author.

Marcos, Texas, or the Bell County Museum in Belton, Texas. The museum offers excellent exhibits about the Gault Site, as well as on the more recent history of the area.

At the site, the surface of the Central Texas Uplift is underlain by Upper Cretaceous marine carbonate rocks. The stratigraphically highest (youngest) formation is known as the Georgetown Limestone, a fractured and permeable unit. The Edwards Limestone, beneath the Georgetown, is cavernous and locally highly permeable and contains abundant chert and flint nodules. The Comanche Peak Limestone, below the Edwards, is a tight formation that does not transmit water readily and generally retards downward movement of groundwater. A small stream valley, eroded into these formations, has been cut below the water table near the base of the Edwards Limestone, thus producing the conditions for springs to issue from the Edwards aquifer here.

The uplands adjacent to the site are generally dry with low-growing oaks and junipers, but, because of the excellent year-round water supply, the site itself is an oasis of trees, grasses, and forbs (broadleaf herbs) usually found in regions with a cooler and wetter climate. Any people exploring the area would have been drawn to this sheltered, well-watered place with abundant plant, animal, and tool-making resources.

Archaeological work has revealed that peoples came to the site on and off for more than 14,000 years, taking advantage of its natural resources. They quarried chert and flint at least from Clovis Culture times (13,500–12,900 years ago) onward. From 1929, when the site was discovered, until 1990, artifact collectors dug up large areas down through Archaic-age layers of sediment searching for artifacts, but generally they did not reach the more deeply buried Paleoindian artifacts. More than two million of these older artifacts have subsequently been found and preserved. At one test pit, pre-Clovis artifacts date from about 14,500 years ago, a date that fits in with the few other pre-Clovis sites so far authenticated in the Americas. Data from Gault seem to indicate clearly that Clovis peoples lived here in a settlement, rather than following the nomadic lifestyle usually attributed to them.

Balcones Fault Zone

G

In south central Texas, the Balcones Fault Zone forms the border between the Great Plains and the Coastal Plain to the east. Covert Park at Mount Bonnell on the western side of Austin, Texas, offers a spectacular view of the relief along this part of the border between the two physiographic provinces. The mount, an erosional remnant of Cretaceous limestone

Fig. 72. A view of part of Lake Austin and the change in slope on the Balcones Fault Zone (*upper right*) near Austin, Texas. This is another example of a structural boundary between adjacent physiographic provinces, in this case the Great Plains (*right*) and Gulf of Mexico Coastal Plain (*left*) Provinces. Photo by author.

formations, is on the uplifted side of the Balcones Fault Zone. Part of the Edwards Limestone is exposed at the top, underlain by the Walnut and Glen Rose Formations. You can see sweeping views of the city, the uplifted block to the south, and parts of Lake Austin, a reservoir on the Canadian River below the park. You can also walk on trails on natural and man-made exposures of Edwards Limestone.

SITE 56

Confluence of Pecos River and Rio Grande

G

I included a small part of Mexico in the Great Plains, following boundaries on geologic maps that cross the border. Initially I did not know by my own observation if the landforms, geology, and vegetation continued from Texas across the Rio Grande or if they changed abruptly on the Mexican side of the river.

Fig. 73. The mouth of the Pecos River at its confluence with the Rio Grande in Texas (*right to left just above center*). Mexico is in the background. Note that rock colors and vegetation are similar on both sides of the Rio Grande. The rocks are Upper Cretaceous marine sedimentary formations. Photo by author.

Given the unsettled situation along the border, I did not want to drive into Mexico to see the area.

Fortunately, I found a way to get a view. On the east side of the Pecos River where U.S. Highway 90 crosses the river, a road off to the south takes you to picnic areas and a boat landing that are quite close to the confluence of the Pecos with the Rio Grande. There you can see the rocks, vegetation, and landforms on both sides of the Rio Grande. Cretaceous limestone beds crop out on both sides of the river without apparent disruption. The vegetation and landscape look the same. While these observations are not proof positive that the Edwards Plateau Section continues across the Rio Grande, they are suggestive that the views of some Mexican and U.S. researchers are correct. That is good enough for me.

SITE 57

Seminole Canyon State Park

G, P, A

I may remember this Texas park for, among other things, being the place where I saw my first tarantula in the wild. It was, big, fast, and aggressive!

Seminole Canyon is an intermittent stream valley eroded into the Lower Cretaceous Devils River Limestone, roughly equivalent in age to the Georgetown Limestone that occurs in the vicinity of the Gault Site (site 54). The river in the canyon is a tributary of the Rio Grande. Soils here are generally thin,

Fig. 74. Fate Bell Shelter, Seminole Canyon State Park, Texas. The rocks were undercut to form the shelter by the stream that formed Seminole Canyon during its erosional development. Photo by author.

Fig. 75. Colored pictographs at Fate Bell Shelter, Seminole Canyon State Park. A dry climate, overhanging rocks, and an isolated location contributed to preservation of this rock art. Photo by author.

and vegetation is sparse and xerophytic, with many spiny plant species. As erosion of the canyon proceeded, the river meandered back and forth, undercutting the limestone and forming what Native American peoples of the Archaic Culture used as rock shelters. One, the Fate Bell Shelter, found in the park area, may only be visited with a park guide at certain times of the day and the year. Several miles of hiking trails on the upland south of the park headquarters offer outstanding views of the canyon and the Rio Grande Valley.

The Fate Bell Shelter has many colored pictographs in the Lower Pecos River Style believed to have been made by Middle Archaic people about 4,000 years ago. The murals in black, red, orange, and yellow are absolutely spectacular.

Afterword

Seminole Canyon and its pictographs! What a place to end a tour of geological sites of the Great Plains of North America. I have enjoyed visiting all of these places over the years. I especially enjoyed visiting the many new ones in Alberta, Saskatchewan, Colorado, New Mexico, and Texas in preparation

Fig. 76. An iconic image of the Great Plains as they are known in the popular imagination. This area is part of the glaciated Missouri Plateau north of Great Falls, Montana. Meriwether Lewis described the land here as having "somewhat the appearance of an ocean" (July 17, 1806). These grasslands are directly underlain by a Late Wisconsinan ground moraine, sediment left behind when the last Late Pleistocene ice sheet melted away. Photo by author.

for writing this book. Of course, many of the sites described do not fit the stereotypical view of the Great Plains as continuously flat and grass-covered, but that was one of my goals. I hope that I have provided you with a better understanding of the tremendous variation in the landscapes, geology, ecology, and archaeology of this wondrous place and have given you the itch to explore it either for the first time or once again in more detail.

Geologic Subdivisions of the Great Plains

Geologists have divided the Great Plains Province into ten subdivisions called sections (fig. 4). The boundaries between sections are sometimes fairly sharply defined by a structural element like a major fault or fold, but most have boundaries that are more arbitrary. I am in general agreement with most and do think that each can be separated from the others. These include the Alberta Plain (1); the Missouri Plateau (2), further subdivided into glaciated (2a) and unglaciated (2b) parts; the Black Hills (3) in South Dakota and Wyoming; the High Plains (4) in parts of South Dakota, Wyoming, Nebraska, Colorado, Kansas, Oklahoma, New Mexico, and western Texas; the Plains Border (5) in Nebraska, Kansas, western Oklahoma and Texas; the Colorado Piedmont (6); the Raton (7) in northeastern New Mexico and southeastern Colorado; the Pecos Valley (8) in New Mexico and western Texas; the Edwards Plateau (9) in southern Texas and northeastern Mexico; and the Central Texas Uplift (10).

1. Alberta Plain

Our Canadian friends call their part of the Great Plains the Alberta Plain (fig. 4:1), an area extending southeastward to the international boundary. One might quibble with using a cultural feature like the Canadian-U.S. border to define a physiographic region, but the border is roughly at the drainage divide between the Missouri River basin in the United States and Canada, and so the subdivision is as justifiable as most. I have, however, drawn the southern boundary generally along the northern edge of the Missouri River drainage basin. This pushes the line into parts of southern Alberta and Saskatchewan and parts of northern Montana and North Dakota, although I have not followed exactly this break in the vicinities of the Cypress Hills and Grasslands National Park in Saskatchewan.

The Alberta Plain is quite distinct and separate from the much higher and structurally far more complex Canadian Rockies to the west and the topographically lower Saskatchewan Plain to the east. The Alberta Plain is mostly at an altitude of about 2,500–3,500 feet but includes higher areas such as the Cypress Hills that have tops reaching altitudes of about 4,700 feet. This relief of more than 2,000 feet appears to have been enough so that the ice sheets in this part of Canada during the several Pleistocene ice ages did not overtop them. The Porcupine Hills south of Calgary are also erosional remnants, but they have less relief.

2. Missouri Plateau

The Missouri Plateau (fig. 4:2a and 2b) extends south from the Canadian-U.S. border through parts of the states of Montana, North Dakota, Wyoming, South Dakota, and Nebraska. The northern and eastern parts of the plateau (section 2a) were covered by ice sheets during the Pleistocene. The region is marked by many distinctive glacial features, including large boulders of called erratics carried from great distances by the moving ice. The remainder of the section (2b) does not have any evidence of having been glaciated during the Pleistocene. The Missouri Plateau is more deeply dissected by the Missouri River and its tributaries than is the Alberta Plain. This is due, at least in part, to greater uplift of the Great Plains and other land in the western United States than in Canada during the last 10 million years, as well as to the erosional and depositional effects of the Pleistocene ice sheets.

Bedrock beneath the surface of the Missouri Plateau is generally composed of nearly horizontal layers of sedimentary rocks covered by variable thicknesses of geologically younger sediments. Both parts of the plateau have some strata either cut and offset by earthquake faults or uplifted and down warped into folds. The uplifts, generally

2,000 to 4,000 feet above the adjacent plains, formed primarily during the Paleogene Period. They include the Sweetgrass Hills, Bearpaw, Little Rocky, Highwood, Moccasin, Judith, Little Belt, and Big Snowy Mountains in Montana and the geologically older, larger, and higher Black Hills of South Dakota and Wyoming.

The western border of this section runs along the break with the Northern Rocky Mountains and parts of the Middle Rocky Mountains and Wyoming Basin physiographic provinces. I have followed some earlier geologists and have not included the areas of intermountain basins such as the Wyoming Basin, Big Horn Basin, and others in the Middle Rocky Mountains that otherwise connect with the Great Plains.

3. Black Hills

The Black Hills Section (fig. 4:3) is an area of the Earth's crust that, principally at the end of the Cretaceous Period, was uplifted by tectonic forces above the level of the adjacent parts of the unglaciated Missouri Plateau on all sides. This dome-shaped mountain has sedimentary rock, including some cavernous Mississippian limestone formations, tilted in all directions away from a geologically older Precambrian central core of igneous and metamorphic rocks.

4. High Plains

The High Plains Section (fig. 4:4) has traditionally been a term applied to an area of the Great Plains stretching from south central South Dakota across parts of southeastern Wyoming, much of Nebraska, parts of eastern Colorado, western Kansas and Oklahoma, western Texas, and eastern New Mexico. This section is generally a topographically higher area and less well dissected by rivers than adjacent ones. It is directly underlain mostly by the Ogallala Group of rocks of Miocene age.

I have included both the Nebraska Sand Hills and the Loess Plains and Hills in the High Plains Section. The Sand Hills generally has sharply demarcated borders that allow visitors to separate it easily from other parts of the High Plains Section on the north, west (fig. 10), and southwest and from the Plains Border Section to the south. Much of the land surface in the Nebraska Sand Hills is today covered with grass. Underneath the vegetation lies wind-deposited Pleistocene and Holocene sand formed into dunes of several kinds, which can be separated from one another on aerial photographs, topographic maps, and satellite images. Beneath these sand deposits are some Pliocene eolian sands and thick river deposits of sand and gravel. The gravel clasts in these deposits are coarser than any geologically earlier deposits beneath the area.

Generally to the east and south of the main Nebraska Sand Hills, Pleistocene and Late Holocene eolian sandy silts and silts (called loess) cover the older Loess Plains and Hills area. In some places the loess deposits are more than 200 feet thick. Here also are some smaller areas covered by sand dunes that formed at the same times as the main part of the Sand Hills north of the Platte and North Platte Rivers.

The boundary of the High Plains extends much farther east in Nebraska than it does in South Dakota and Kansas. I was tempted to separate the Sand Hills and the Loess Plains and Hills into a separate section or two but found that the reasons for keeping the areas in the broader High Plains are geological. That is, the Ogallala Group of rocks generally underlies most of the areas in Nebraska covered by the Sand Hills and the Loess Plains and Hills. Furthermore, the western boundary of the Pleistocene glacial boundary lies just to the east.

5. Plains Border

The Plains Border Section (fig. 4:5) is located in parts of Nebraska, Kansas, Oklahoma, and Texas. It is generally directly underlain by Cretaceous sedimentary rocks that gently tilt to the west. These rocks are covered in many places with Quaternary river- and wind-transported deposits, some of which are rather thick in river valleys and on upland areas.

6. Colorado Piedmont

Entirely in the state of Colorado, the Colorado Piedmont Section (fig. 4:6) is bounded on the north and east by the High Plains, on the south by the Raton Section, and on the west by the Southern

Rocky Mountains. This section is more deeply eroded by rivers than the adjacent High Plains and has greater relief or elevation difference between adjacent higher and lower areas. Particularly as you approach the Rocky Mountains, the relief differences become greater and erosional remnants like Castle Rock in the city of the same name and other buttes and mesas become more prominent. Nearly horizontally layered sedimentary rocks lie beneath these landforms and can be seen exposed in many areas where erosion has been or is active.

7. Raton

The Raton Section (fig. 4:7) lies to the south of the Colorado Piedmont, beginning south of the Arkansas River drainage basin. I have drawn the boundary line in places somewhat south of where others have because I have found that the best break (which can be seen along I-25 in Colorado, for example) is south of Greenhorn Creek, where the land flattens out and is not so deeply eroded.

This section has vast areas covered by dark gray to black, hardened lava flows and volcanic cones. These geologically young Cenozoic volcanic rocks and landforms and other intrusive igneous rocks are exposed at the surface in many areas in the Raton Section but are generally not found on the adjacent High Plains, Colorado Piedmont, or Pecos Valley Sections. These igneous bodies include El Huerfano in Huerfano County and the Spanish Peaks south of La Veta, Colorado.

8. Pecos Valley

The Pecos Valley Section (fig. 4:8) lies south of the Raton Section and west of the High Plains Section. It is underlain by Upper Paleozoic limestone and other ocean deposits. The limestone has been partially dissolved in places beneath the land surface, producing caverns, sinkholes, and disappearing streams.

9. Edwards Plateau

The western part of the Edwards Plateau Section (fig. 4:9) is south of the Pecos Valley and High Plains Sections in western Texas. The land surface of the plateau is directly underlain by Cretaceous marine limestone beds that have also been differentially dissolved in places, forming cavern systems. The eastern side of the plateau is along the Balcones Fault Zone, along which the rocks of the plateau have been uplifted with respect to the adjacent Gulf of Mexico Coastal Plain. The Pecos River and other tributaries of the Rio Grande have eroded steep-walled, deep valleys into the limestone. The layers of limestone usually appear to be nearly horizontal in the Plateau area. This generally structurally uncomplicated plateau and its rocks continue across the Rio Grande into Mexico. For this reason I have included that part of Mexico in the Great Plains.

10. Central Texas Uplift

The final section of the Great Plains, the Central Texas Uplift Section (fig. 4:10), lies to the northeast of the Edwards Plateau. It is bounded on the east by the Balcones Fault Zone, which separates it from the Gulf of Mexico Coastal Plain. The uplift is a dome feature with a central core of igneous and metamorphic rocks surrounded by layers of marine sedimentary rock dipping away from the center, similar to the structure of the Black Hills but geologically much older. The relief of the uplift is not nearly as great as that of the Black Hills. The granites in the central core form large, rounded erosional remnants called monadnocks, which exfoliate or naturally peel off in layers like an onion, due largely to weathering.

Chronology of the Development of Some Geological Concepts

1669 Nicholas Steno's principles of deposition of layered sediments and sedimentary rocks.

1774 A. G. Werner, *On the External Characters of Minerals* (German; first mineral text).

1786 A. G. Werner, *Short Classification and Description of Various Rocks* (German).

1795 James Hutton, *Theory of the Earth with Proofs and Illustrations*.

1796 Richard Kirwan, *Elements of Mineralogy* (2nd edition with rock structures detailed).

1804 Start of Lewis and Clark expedition.

1815 First modern geologic map published by William Smith in England.

1822 Cretaceous System of rocks defined in France.

1824 First report of Cretaceous rocks and fossils in eastern United States.

1833 Start of Prince Maximilian zu Wied-Neuweid and Karl Bodmer's travels along Missouri River.

1837 Louis Agassiz proposes his glacial theory.

1838 Continental glaciation recognized in eastern United States.

1839 Samuel Morton reports Cretaceous fossils from Missouri River basin.

1841 Start of Charles Lyell's travels to parts of North America.

1845 Charles Lyell's U.S. geologic map shows Cretaceous rocks in northeastern Nebraska.

1859 Charles Darwin, *On the Origin of Species by Means of Natural Selection*.

1860 General area glaciated in North America identified.

1877 G. K. Gilbert's observations on rivers and river terraces published.

1895 John Wesley Powell draws first map of U.S. physiographic provinces, including the Great Plains.

1899 Overall distribution of Cretaceous strata in United States shown on maps and noted in reports.

1912 Alfred Wegener proposes theory of "Continental Drift."

1913 Arthur Holmes publishes first geologic timescale based on radiometric dates.

1931 Arthur Holmes first proposes what became known as the theory of plate tectonics.

1960 Theory of plate tectonics rediscovered and begins to be accepted generally.

APPENDIX 3

Cautions for Travelers on the Great Plains

The Great Plains is a wonderful place to visit, but it is generally sparsely settled and can be dangerous. I suggest that any reader planning to visit keep the following cautions always in mind to increase the likelihood of having a safe and enjoyable experience.

- Always carry an adequate supply of water and food in case of breakdowns or other emergencies while you are on the road in remote areas. To me, adequate water means at least a gallon per person per day.
- Always wear a broad-brimmed hat and sunscreen. The dry air, altitude, and sun reflection at all times of the year can lead to significant skin drying and sunburn if your skin is unprotected.
- Always fill up your fuel tank at or shortly after you have used half of its capacity. Fuel stations in the region are sometimes in short supply, and you cannot assume that there will be one available when you need one.
- Always service your vehicle before you depart.
- Always leave an itinerary with appropriate people. If you decide to change any part of the itinerary, notify them.
- Always assume that your cell phone will not work. There are many remote areas with no service.
- Always keep an eye on the weather around you. Storms can develop and move rapidly. These include severe thunderstorms (with accompanying high winds, hail, and tornadoes), dust storms, ice storms, snowstorms, and blizzards. Flash floods can occur anywhere on the Great Plains and may affect roadways and areas

some distance away from where the storm occurred. Never try to drive through flooded or flooding roads. If you are out in the open and away from your vehicle when a storm approaches, seek shelter immediately.

- Always carry a blizzard kit in cold months, even when traveling in usually warm places like the southern parts of the Great Plains. The kit should include water, food, blankets, flashlight, shovel, gallon can(s) for bodily wastes (who wants to go outside in a blizzard?), and a first-aid kit.
- When a sign notes that an unpaved road is "impassable when wet," assume that the writer knows what he or she is talking about and do not try to travel on it when it is wet. If weather conditions begin to deteriorate, get back to a paved road as quickly as possible. If that is not possible, get off the road surface onto somewhat higher, vegetation-covered ground and wait until the roads dry out. Do not assume that four-wheel drive will keep you safe on unpaved roads when they are wet.
- Stay in your vehicle if you are stranded for any reason.
- Always ask permission to walk on private lands.
- Never collect on state or federal lands without a permit. Do not collect on private property or Indian reservations without permission.
- Always use a walking stick for support and balance up and down slopes.
- Be careful of dangerous animals and plants. Plants include poison ivy, cactus, and other plants with sharp spines. Animals include mosquitoes, ants, other biting and stinging insects, tarantulas and scorpions in the south (scorpions range into parts of south central Nebraska), other poisonous spiders, and poisonous snakes such as rattlesnakes (found at least as far north as southern Alberta and Saskatchewan). Livestock may also be dangerous.
- If you plan on visiting a museum or a park, telephone to check on the days and times when it is open. Do not count on information obtained from the web to be up-to-date. Sometimes websites have not been updated in years.

absolute dating The calculation of age in years before the present era (BP), usually on the basis of radioactive isotopes.

alluvium A general term for clay, silt, sand, or gravel deposited by a stream or other running body of water.

anticline A fold, generally convex upward, whose core contains stratigraphically older rocks.

arkose A sandstone, commonly coarse grained and pink or reddish, typically composed of angular grains derived from the rapid disintegration of granitic rocks, and often resembling granite.

atmosphere The mixture of gases that surrounds the Earth, being held thereto by gravity.

badlands Stream-dissected topography developed on surfaces with little or no vegetative cover on unconsolidated clays or silts.

basalt A dark-colored, fine-grained igneous rock.

bedrock A general term for the rock, usually solid, that underlies soil or other unconsolidated surface material.

biosphere All living organisms of the Earth and its atmosphere.

biota All living organisms of an area; the flora and fauna considered as a unit.

calcareous Said of a substance that contains calcium carbonate.

caldera A large, basin-shaped volcanic depression formed by collapse during an eruption.

caliche A reddish-brown to buff or white calcareous material found in layers on or near the surface of stony soils in arid and semi-arid regions. It is composed largely of soluble calcium carbonate together with gravel, sand, silt, and clay.

cascade Water descending over steeply slanting rocks; a shortened rapids.

chert A hard sedimentary rock consisting mostly of very small interlocking quartz crystals, which fractures like glass. It may be white or many other colors depending on trace elements included.

clast An individual constituent, grain, or fragment of a sediment or rock produced by the mechanical or chemical disintegration of a larger rock mass.

coastal plain A low, generally broad plain that has its margin on an oceanic shore and its strata either horizontal or very gently sloping toward the water and that generally represents a strip of recently emerged sea floor.

concretion A hard, compact mass of mineral matter, normally spherical in shape, formed by precipitation from aqueous solution about a nucleus or center, such as a leaf, shell, or bone, in a sedimentary or a fragmented volcanic rock.

conglomerate A coarse-grained clastic sedimentary rock, composed of rounded fragments larger than two millimeters in diameter containing fine-grained particles in the interstices and commonly cemented by calcium carbonate, iron oxide, silica, or hardened clay.

core The central zone or nucleus of the Earth's interior surrounded by the mantle.

craton A part of the Earth's crust that has attained stability and has been little deformed for a prolonged period.

cross-bed A single bed, inclined at an angle to the main planes of stratification. Found in sedimentary structures.

crust The outermost layer or shell of the Earth, surrounding the mantle.

crustal rebound The upward flexing of the Earth's crust, commonly as the result of release of pressure or weight, such as from the melting of an ice sheet.

cuesta A hill or ridge with a gentle slope on one side and a steep slope on the other.

deformation A general term for the process of folding, faulting, or shearing of rocks as a result of Earth stresses.

dike A tabular intrusion of igneous rock that cuts across the bedding of existing rocks.

dome A general term for any smoothly rounded landform or rock mass that resembles the dome of a building. Also applied broadly to up-arched regions.

entrenched meander A meander carved downward into the surface of a valley, preserving its original pattern with little modification.

eolian Pertaining to the wind; especially said of such deposits as loess and dune sand, or of erosion and deposition accomplished by the wind.

erosion The general process whereby the materials of the Earth's crust, such as soil and rock, are loosened, dissolved, or worn away and moved from one place to another by natural agents, including running water (including rainfall), waves and currents, moving ice, or wind.

erosional remnant A topographic feature that remains or is left standing above the general land surface after erosion has reduced the surrounding area, for example, a monadnock, butte, mesa, or stack.

erratic A large rock fragment carried by glacial ice and deposited at some distance from the outcrop from which it was derived, generally resting on sediment or rock of a different kind.

escarpment (a) A long, generally continuous cliff or steep slope facing in one general direction, breaking the continuity of the land by separating two surfaces; produced by erosion or faulting. (b) A steep, abrupt face of rock, often the highest strata in a line of cliffs, generally marking the outcrop of a resistant layer occurring among softer strata.

fault A discrete surface or zone of surfaces separating two rock masses across which one mass has slid past the other.

fission tracks The paths of radiation damage made by nuclear particles in a mineral or glass by the spontaneous fission of uranium-238. Fission-track dating is a means of absolute dating whereby ages in years are calculated by determining the ratio of spontaneous fission tracks to induced ones.

fluvial Produced by the action of a river or stream.

formation A persistent body of rock with recognizable boundaries and common characteristics that is mappable at the Earth's surface or traceable in the subsurface. It is generally, but not necessarily, either tabular (book shaped) or lens shaped.

geologic map A map that displays information such as the distribution, nature, and age relationships of rock units and the occurrence of structures (folds, faults, joints), mineral deposits, and fossil localities.

graben An elongate depression or basin bounded on at least two sides by faults.

group A sequence of two or more contiguous formations with significant features in common.

hogback Any ridge with a sharp summit and steep slopes of nearly equal inclination on both flanks and resembling in outline the back of a hog.

hydrosphere The waters of the Earth, as distinguished from the rocks, living things, and air.

igneous Said of a rock or mineral that solidified from molten or partly molten material; also applied to processes leading to, related to, or resulting from the formation of such rocks.

intrusion The process of emplacement of magma in preexisting rock; also, the igneous rock so formed within the surrounding rock.

isotope One of two or more species of the same chemical element having the same number of protons in the nucleus but a different number of neutrons.

joint A crack or fracture in rock not accompanied by movement or dislocation.

lens A geologic deposit bounded by converging surfaces (at least one of which is curved), thick in the middle and thinning out toward the edges, resembling a convex lens.

lithosphere The Earth's crust and upper part of the underlying mantle that behave relatively rigidly and can bend or flex when Earth forces are applied.

loess Windblown dust of Pleistocene and older geologic ages, carried from desert surfaces, alluvial valleys, or glacial deposits. A widespread, homogeneous blanket deposit, it covers extensive areas from north central Europe to eastern China as well as the north central and Pacific northwestern sections of the United States. Generally buff to light yellow to yellowish-brown, it erodes in steep or nearly vertical faces. Pronounced *luhss*.

magma Naturally occurring molten or partially molten rock material, generated within the Earth.

mantle The zone of the Earth below the crust and above the core.

meander One of a series of regular, freely developing sinuous curves, bends, loops, turns, or windings in the course of a stream.

metamorphic Said of rocks changed mineralogically, chemically, or structurally by conditions at depth, below the surface zones of weathering and cementation, and that differ from the conditions under which the rock originated.

mudstone A sedimentary rock composed mostly of very small particles, less than two millimeters in diameter, with included larger particles.

normal fault A fault in which the block resting on top of the angling fault plane slides down relative to the block below.

outcrop That part of a geologic formation or structure that appears at the surface of the Earth.

paleosol A soil that formed on the landscape in the past, now often buried by subsequent deposition.

pegmatite An exceptionally coarse-grained igneous rock with most grains one or more centimeters in diameter. Its mineral composition is usually, but not necessarily, that of granite.

physiographic Pertaining to the form of the Earth's surface features.

plateau Any comparatively flat uplifted area of great extent and elevation; often dissected by deep valleys.

quartzite A metamorphic rock consisting mainly of the mineral quartz and formed from recrystallization of sandstone.

relative dating The proper chronological placement of a feature or event in the geologic timescale without reference to its absolute age.

relief The vertical difference in elevation between the tops or summits of highlands or mountains and the lowlands or valleys in a given place. A geologic feature showing great variation in elevation has "high relief," and one showing little variation has "low relief."

reverse fault A fault in which the block above (resting on) the fault plane moves up relative to the block below.

sandstone Any sedimentary rock composed of an agglutination of grains of sand.

schist A crystalline metamorphic rock that can readily be split into thin flakes or tabular slabs because most of its crystals are oriented in parallel planes. The mineral composition is not an essential factor in its definition unless specifically included in the rock name.

sedimentary Said of rock resulting from the consolidation of loose, fragmented material that has accumulated in layers. It includes materials precipitated from water (like salts) and also organically produced solids (coals and some limestones).

sill A tabular igneous intrusion that parallels the bedding plane of a sedimentary rock formation or the foliation of metamorphic rock.

siltstone A sedimentary rock composed mainly of an agglutination of grains of silt particles ranging in diameter from 0.00016 and 0.0025 inches.

stock An igneous intrusion.

strata Tabular or sheet-like layers of sedimentary rock, visually separable from other layers above and below.

stratigraphic succession A chronological sequence of strata, from older below to younger above.

supervolcano Any volcano capable of producing an explosive volcanic eruption with an ejecta mass greater than 10^{15} kilograms. These include Yellowstone Caldera in the United States and Mount Taupo in New Zealand.

tableland A plateau bordered by abrupt, cliff-like edges rising sharply from the surrounding lowland.

talus A slope formed by an accumulation of rock debris; rock debris at the base of a cliff.

tectonic Of or relating to the deformation of the Earth's crust, the forces involved in or producing such deformation, and the resulting forms.

till Unsorted and unstratified materials deposited directly by and underneath a glacier, consisting of a heterogeneous mixture of clay, silt, sand, gravel, and boulders ranging widely in size and shape.

topographic map A map showing horizontal and vertical positions of the features being represented, with the relief being shown in measurable form, commonly by means of contour lines. It is generally in sufficiently large scale to show selected man-made and natural features, including rural buildings, vegetation, roads, and drainages.

tuff A consolidated or cemented volcanic ash and coarser volcanic debris.

unconformity A substantial break or gap in the geologic record where a rock unit is overlain by another that is not next in stratigraphic succession. It results from a change that caused deposition to cease for a considerable span of time, and it normally implies uplift and erosion with loss of the previously formed record.

uplift A structurally high area in the Earth's crust, produced by positive movements that raise or upthrust the rocks, as in a dome or arch.

xerophyte A plant adapted to dry conditions; a desert plant.

BIBLIOGRAPHY

Agenbroad, L. D. and J. I. Mead, eds. *The Hot Springs Mammoth Site: A Decade of Field and Laboratory Research in Paleontology, Geology, and Paleoecology*. Hot Springs SD: Mammoth Site of South Dakota, 1994.

Alliance, Synthetic-Aperture Radar Imagery X-Band, Alliance Quadrangle. Map. 1 Sheet. Original Scale 1:250,000. U.S. Geological Survey, 1988.

Atwood, W. W. *The Physiographic Provinces of North America*. Boston: Ginn and Company, 1940.

Brantley, S. R. *Volcanoes of the United States*. Pamphlet. U.S. Geological Survey, 1995.

Carter, J. M., D. G. Driscoll, J. E. Williamson, and V. A. Lindquist. *Atlas of Water Resources in the Black Hills Area, South Dakota*. Hydrologic Investigations Atlas HA-747. U.S. Geological Survey, 2002.

Darton, N. H. and S. Paige. "Geologic Atlas of the United States, Central Black Hills Folio, South Dakota." *U.S. Geological Survey Folio* 219 (1925): 1–34.

Diffendal, R. F., Jr. "Earth in Four Dimensions—Development of Ideas of Geologic Time and History." *Nebraska History* 80, no. 3 (1999): 95–104.

———. "Plate Tectonics, Space, Geologic Time and the Great Plains—A Primer for Non-geologists." *Great Plains Quarterly* 11, no.1 (1991): 83–102.

Diffendal, R. F., Jr. and A. P. Diffendal. *Lewis and Clark and the Geology of the Great Plains*. Educational Circular 17. Lincoln: Conservation and Survey Division, University of Nebraska, 2003.

Douglas, R. J. W., ed. *Geology and Economic Minerals of Canada*.

Economic Geology Report no. 1. Geological Survey of Canada, 1970. 12 maps and charts in separate box.

Fenneman, N. M. *Physiography of Western United States*. New York: McGraw-Hill, 1931.

Gerlach, A. C., ed. *The National Atlas of the United States of America*. Washington DC: U.S. Geological Survey, 1970.

Gill, J. R., and W. A. Cobban. *Stratigraphy and Geologic History of the Montana Group and Equivalent Rocks, Montana, Wyoming, and North and South Dakota*. Professional Paper 776. U.S. Geological Survey, 1973.

Hills, R. C. "Geologic Atlas of the United States, Walsenburg Folio, Colorado." *U.S. Geological Survey Folio* 68 (1900): 1–6.

Hoffman, P. F. "United Plates of America, the Birth of a Craton: Early Proterozoic Assembly and Growth of Laurentia." *Annual Reviews of Earth and Planetary Science* 16 (1988): 543–603.

Hunt, C. B. *Natural Regions of the United States and Canada*. San Francisco: W. H. Freeman, 1974.

Johnson, K., and R. Troll. *Cruisin' the Fossil Freeway*. Golden CO: Fulcrum, 2007.

Jones, C. H., G. L. Farmer, B. Sageman, and S. Zhong. "Hydrodynamic Mechanism for the Laramide Orogeny." *Geosphere* 7 no. 1 (2011): 183–201.

Jones, C. H., K. H. Mahan, L. A. Butcher, W. B. Levandowski, and G. L. Farmer. "Continental Uplift through Crustal Hydration." *Geology* 43, no. 4 (2015): 355–58.

Kauffman, E. G. "Geological and Biological Overview: Western Interior Cretaceous Basin." *Mountain Geologist* 14, no. 3–4 (1977): 75–99.

Kaul, R. B., and S. B. Rolfsmeier. *Native Vegetation of Nebraska*. Map. 1 sheet. Scale 1:1,000,000. Lincoln: Conservation and Survey Division, University of Nebraska, 1993.

Kolbert, E. *The Sixth Extinction*. New York: Henry Holt, 2014.

Korus, J. T., L. M. Howard, A. R. Young, D. P. Divine, M. E. Burbach, J. M Jess, and D. R. Hallum. *The Groundwater Atlas of Nebraska*. 3rd rev. ed. Resource Atlas 4b. Lincoln: Conservation and Survey Division, University of Nebraska, 2013.

Lavin, S. J., F. M. Shelly, and J. C. Archer. *Atlas of the Great Plains*. Lincoln: University of Nebraska Press, 2011.

Lee, W. T. "Geologic Atlas of the United States, Raton—Brilliant—

Koehler Folio, New Mexico-Colorado." *U.S. Geological Survey Folio* 214 (1922): 1–17.

Lewis, C. *The Dating Game: One Man's Search for the Age of the Earth.* New York: Cambridge University Press, 2000.

Light Brown, C. *Geology of the Great Plains and Mountain West.* White River Junction VT: Nomad Press, 2011.

Love, J. D. "Cenozoic Sedimentation and Crustal Movement in Wyoming." *American Journal of Science, Bradley Volume,* 258-A (1960): 204–14.

Lovelock, J. E. "Gaia as Seen through the Atmosphere." *Atmospheric Environment* 6, no. 8 (1972): 579–80.

———. *Gaia: A New Look at Life on Earth.* 5th ed. Oxford: Oxford University Press, 2000.

Marshak, S. *Earth: Portrait of a Planet.* 2nd ed. New York: W. W. Norton, 2005.

McPhee, J. *Rising from the Plains.* New York: Farrar, Straus, Giroux, 1986.

Miller, S. "Fossils and Geology of the Greenhorn Cyclothem in the Comanche National Grasslands, Colorado." *Geological Society of America Field Guide* 33 (2013): 269–78.

Nereson, A., J. Stroud, K. Karlstrom, M. Heizler, and W. McIntosh. "Dynamic Topography of the Western Great Plains: Geomorphic and 40Ar/39Ar Evidence for Mantle-Driven Uplift Associated with the Jemez Lineament of NE New Mexico and SE Colorado." *Geosphere* 9, no. 3 (2013): 521–45.

Neuendorf, K. K. E., J. P. Mehl Jr., and J. A. Jackson. *Glossary of Geology.* 5th ed. Alexandria VA: American Geological Institute, 2005.

Newton, H., and Jenney, W. P. *Report of the Geology and Resources of the Black Hills of Dakota.* Washington DC: Government Printing Office, 1880.

Oldroyd, D. "Maps as Pictures and Diagrams: The Early Development of Geological Maps." In *Rethinking the Fabric of Geology,* edited by V. R. Baker, 41–101. Special Paper 502. Boulder CO: Geological Society of America, 2013.

Powell, J. W. "Physiographic Regions of the United States." *National Geographic Monographs* 1, no. 3 (1895): 65–100.

Price, L. *The Geology of Northern New Mexico's Parks, Monuments, and Public Lands.* Socorro: New Mexico Bureau of Geology and Mineral Resources, 2010.

Renne, P. R., C. J. Sprain, M. A. Richards, S. Self, L. Vanderkluysen, and K. Pande. "State Shift in Deccan Volcanism at the Cretaceous-Paleogene Boundary, Possibly Induced by Impact." *Science* 350, no. 6256 (2015): 76–78.

Richardson, G. B. "Geologic Atlas of the United States, Castle Rock Folio." *U.S. Geological Survey Folio* 198 (1915): 1–13.

Rossum, S., and S. J. Lavin. "Where Are the Great Plains? A Cartographic Analysis." *Professional Geographer* 52 (2000): 543–52.

Roy, M., P. U. Clark, R. W. Barendregt, J. R. Glasmann, and R. J. Enkin. "Glacial Stratigraphy and Paleomagnetism of Late Cenozoic Deposits of the North-Central United States." *Geological Society of America Bulletin* 116, no.1–2 (2004): 30–41.

Skelton, P., ed. *The Cretaceous World*. Cambridge: Cambridge University Press, 2003.

Stein, C. A., J. K. Kley, S. Stein, D. Hindl, and G. R. Keller. "North America's Midcontinent Rift: When Rift Met LIP." *Geosphere* 11, no. 5 (2015): 1607–16.

Swinehart, J. B. "Wind-Blown Deposits." In *An Atlas of the Sand Hills*, edited by A. S. Bleed and C. Flowerday, 43–56. Resource Atlas 5b. Lincoln: Conservation and Survey Division, University of Nebraska–Lincoln, 1998.

Swinehart, J. B. and R. F. Diffendal Jr. "Geology of the Pre-dune Strata." In *An Atlas of the Sand Hills*, edited by A. S. Bleed and C. Flowerday, 29–42. Resource Atlas 5b. Lincoln: Conservation and Survey Division, University of Nebraska–Lincoln, 1998.

Swinehart, J. B., V. L. Souders, H. M. DeGraw, and R. F. Diffendal Jr. "Cenozoic Paleogeography of Western Nebraska." In *Cenezoic Paleogeography of the West-Central United States: Rocky Mountain Paleogeography Symposium 3*, edited by R. Flores and S. Kaplan, 209–29. Denver: Society of Economic Paleontologists and Mineralogists, Rocky Mountain Section, 1985.

Thornbury, W. D. *Regional Geomorphology of the United States*. New York: John Wiley & Sons, 1965.

The Top 50 Ecotourism Sites in the Great Plains. Map. 1 sheet. Scale approximately 1:3,168,000. Lincoln: University of Nebraska, Center for Great Plains Studies, 2012.

Trimble, D. E. "The Geologic Story of the Great Plains." *U.S. Geological Survey Bulletin* 14, no. 93 (1980): 1–20.

Tucker, S. T., R. E. Otto, R. M. Joeckel, and M. R. Voorhies. "The Geology and Paleontology of Ashfall Fossil Beds, a Late Miocene (Clarendonian) Mass-Death Assemblage, Antelope County and Adjacent Knox County, Nebraska, USA." In *Geologic Field Trips along the Boundary between the Central Lowland and the Great Plains*, edited by J. T. Korus, 1–22. Field Guide 36. Boulder CO: Geological Society of America, 2014.

Voorhies, M., R. Otto, S. Mosel, and S. Tucker. *Ashfall Fossil Beds State Historical Park and National Natural Landmark: Present View of an Ancient Past*. Lincoln: Board of Regents of the University of Nebraska, 2015.

Weeks, J. B., and E. D. Gutentag. "Bedrock Geology, Altitude of Base, and 1980 Saturated Thickness of the High Plains Aquifer in Parts of Colorado, Kansas, Nebraska, New Mexico, Oklahoma, South Dakota, Texas, and Wyoming." Map. 2 sheets. Scale 1:2,500,000. Hydrologic Investigations Atlas HA-648. U.S. Geological Survey, 1981.

Winchester, S. *The Map That Changed the World: William Smith and the Birth of Modern Geology*. New York: HarperCollins, 2001.

Wishart, D. J. *Encyclopedia of the Great Plains*. Lincoln: University of Nebraska Press, 2004.

Young, G. M. "Evolution of Earth's Climatic System: Evidence from Ice Ages, Isotopes, and Impacts." *GSA Today* 23, no. 10 (2013): 4–10.

INDEX

Page numbers in italics signify graphics.

IN THE DISCOVER THE GREAT PLAINS SERIES

Great Plains Geology
R. F. Diffendal Jr.

Great Plains Indians
David J. Wishart

Discover the Great Plains, a series from the Center for Great
Plains Studies and the University of Nebraska Press, offers
concise introductions to the natural wonders, diverse cultures,
history, and contemporary life of the Great Plains. To order
or obtain more information on these or other University of
Nebraska Press titles, visit nebraskapress.unl.edu.